UNCERTAINTY ANALYSIS, LOADS, AND SAFETY IN STRUCTURAL ENGINEERING

PRENTICE-HALL CIVIL ENGINEERING AND ENGINEERING MECHANICS SERIES

N.M. NEWMARK and W.J. HALL, *Editors*

UNCERTAINTY ANALYSIS, LOADS, AND SAFETY IN STRUCTURAL ENGINEERING

Gary C. Hart

Mechanics and Structures Department
School of Engineering and Applied Science
University of California, Los Angeles

PRENTICE-HALL, Inc., Englewood Cliffs, New Jersey   07632

Library of Congress Cataloging in Publication Data

HART, GARY C.
    Uncertainty analysis, loads, and safety
in structural engineering.

    Includes bibliographies and index.
    1. Structural engineering.   2. Structural
stability.   3. Safety factor in engineering.
I. Title.
TA633.H37   624.1′71   81-8483
ISBN 0-13-935619-3   AACR2

Editorial production supervision
  and interior design by Karen Wagstaff
Manufacturing buyer: Joyce Levatino

Printed in the United States of America

10 9 8 7 6 5 4 3 2 1

ISBN 0-13-935619-3

Prentice-Hall International, Inc., *London*
Prentice-Hall of Australia Pty. Limited, *Sydney*
Prentice-Hall of Canada, Ltd., *Toronto*
Prentice-Hall of India Private Limited, *New Delhi*
Prentice-Hall of Japan, Inc., *Tokyo*
Prentice-Hall of Southeast Asia Pte. Ltd., *Singapore*
Whitehall Books Limited, *Wellington, New Zealand*

To Marianne, Kristine, and my father.

Contents

3   STRUCTURAL ANALYSIS INCORPORATING UNCERTAINTY 55

4   STRUCTURAL SAFETY 104

5   STRUCTURAL LOADS 135

Preface

Structural engineering books and associated university courses have historically focused on the principles of structural mechanics, structural analysis, and structural design. Almost without exception these books assume that the material properties, geometric dimensions, and loading for the structure are exactly known.

This book is intended to be a supplement to existing structural engineering books. While the need for a good foundation in structural mechanics and structural design is as important today as it ever was, it is also fundamentally important that today's structural engineering students and practicing structural engineers recognize that uncertainty exists and learn to evaluate rationally and incorporate it into structural design practice.

The prerequisites for the reading of this book are a strength of materials course and a basic course in statically determinate structures. No prior course in probability or statistics is required. Also, the reader is not expected to have had a course in structural dynamics or vibration theory. It is desirable for the reader to have taken a course in structural design (either concrete, steel, or masonry) in order to appreciate better the material being presented.

The book was written for two audiences. First, senior level students possess the necessary background to be able to cover the material presented here in one college semester. Second, the number of example problems is quite large so that practicing structural engineers can independently read the text and verify their understanding through problem solutions.

Many persons have indirectly contributed to the development of this book. Fellow engineers whom I particularly wish to acknowledge are: Jack R. Benjamin,

John A. Blume, Jon D. Collins, Robert E. Englekirk, Roberto Del Tosto, John H. Wiggins, and J. T. P. Yao.

Many of the ideas presented herein were the result of my research funded by the National Science Foundation and encouraged by Mike Gaus and Jack Scalzi of NSF.

Many students have made valuable suggestions. I am especially grateful to two of my former students, Dr. Sampson Huang and Dr. Marshall Lew.

The typing, retyping, etc., of the text for the more than five years of the book's development were done patiently and professionally by Debby Haines.

GARY C. HART

UNCERTAINTY ANALYSIS, LOADS, AND SAFETY IN STRUCTURAL ENGINEERING

1

Loads and Safety

1-1 THE ROLE OF THE STRUCTURAL ENGINEER

Structural engineering is a complex and demanding profession, requiring the engineer to be businesslike to obtain a new job; a manager to organize and review the work of subordinates; a psychologist to deal with a multitude of different personalities (architects, city planners, building officials, bankers, and at times perhaps a partner). The structural engineer also has the additional concern of "professional responsibility," which refers to the continuous updating of building standards or codes because of new research findings and the direct implementation of such information for special structures that necessitate a more in-depth study than that prescribed by existing standards or codes.

Persons not familiar with structural engineering, and sometimes recently graduated engineers, often do not appreciate the complexity of the role of the structural engineer. Lacking an understanding of the whole role, an individual can easily be overly critical of the engineer's efforts applying to any one aspect of the total role. The reader may wonder what these comments have to do with the subject of this book—uncertainty analysis, loads, and safety in structural engineering. To appreciate our ultimate objective, one must appreciate the role of the structural engineer. It is our objective in this book to provide a vehicle for the continued incorporation into structural engineering of a more rational treatment of loading and strength uncertainties. By such incorporation it becomes possible to discuss more rationally the safety of structures and, perhaps more importantly, provide useful input into the overall economic and social planning of structures.

No technical book can provide the answers to all problems. However, a book can broaden a reader's horizon and establish a foundation for further development of the ideas presented. Properly written, a technical book can also help to eliminate unjustified fear of its subject matter, a fear often arising due to the lack of subject cohesiveness one experiences when reading disjointed technical journal papers or technical conference proceedings.

The author's primary goals will be met if the reader, after an initial inspection, is encouraged to read the book and, after reading it, feels that the material is "not so difficult" and "makes sense." Each chapter contains examples which are based on real structural engineering problems that are intended to provide a bridge between the theory of probability, statistics, and applied mathematics, and actual structural engineering problems.

Is there an immediate financial reward for reading this book? The answer is, perhaps not. However, the book will provide the reader with a better appreciation for the level of uncertainty which exists in both structural modeling and load description, and does also indicate how these uncertainties can be incorporated into designs or code policy-making deliberations. Since new codes are moving in a direction which requires a quantification of such uncertainties, a financial return will most certainly be realized eventually.

1-2 ORGANIZATION OF THE BOOK

The book is divided into a series of chapters. Each chapter starts with an introduction and concludes with references. The introduction explains the purpose of the chapter, and the references provide a selected listing of additional reading material.

Chapter 2 presents all of the probability and statistics background required for this book. The definitions and concepts are illustrated by using structural engineering problems. The reader will, it is hoped, identify many problems which are familiar from previous readings in strength of materials and structural design. The example problems are intended to provide a "break" from the formal definitions of terms in the text and, more important, to provide a real appreciation for the degree of uncertainty inherent in most structural parameters.

Chapter 3 is a "fun" chapter! It shows how the material presented in Chapter 2 can be combined with the reader's background to produce something not only interesting but useful—a rare combination to be sure. Historically, the analysis of a structure involves the selection of the analysis approach, followed by the selection of unique values for the parameters of the model. This chapter describes how the uncertainty in the values of the parameters of the model can be directly accounted for and how the resulting uncertainty in the calculated response quantity can be quantified.

Safety implies failure, and it is becoming more important for structural engineers to not only be more precise in describing what they mean by failure but also to quantify the chance, or probability, that failure will occur. Chapter 4 discusses structural safety and describes approaches which can be used in its evaluation.

Chapter 5 presents an introduction to the loads that are most common in structural engineering. These loads have been studied for many years and there exist many possible approaches for describing them. The complexity of these alternate approaches varies from simple algebraic formulas to highly complex analytical ones. Through the additional reading section at the end of the chapter the reader is provided with a list of additional references citing many of these approaches. It is not our intent in this chapter to summarize all existing approaches; rather, the intent is to show that there is a common strand that can be employed to quantify structural loading.

2

Uncertainty in Structural Engineering

2-1 INTRODUCTION

This chapter presents all of the necessary background in probability and statistics needed for reading this book. The topics of probability and statistics are familiar to engineers and scientists, and there are many books which provide a more comprehensive treatment. Several of these books are listed in the references at the end of this chapter.

The examples in this chapter involve real problems, and the data are representative of the level of uncertainty one can expect in structural engineering.

2-2 SAMPLE SPACE, EVENTS, AND RANDOM VARIABLES

The concept of a random variable is closely related to the conducting of an experiment. If an experiment is performed repeatedly (with all conditions maintained as precisely as possible) and the measured results are identical, then the items which are measured are said to be *deterministic*. However, if the numerical results vary, the items are *random*. In structural engineering, no numerical measurements when carried to four or five significant figures of accuracy, and repeated, will have identical numerical results. Therefore, it is more realistic to view any item measured in an experiment as a random variable, and then to determine whether the observed variations are large enough for concern in regard to the engineering problem under consideration.

In an experiment, any numerically valued item derived from the experiment can be a *random variable*. All possible observations of the random variable comprise its *sample space*. An *event* is a collection, or subset, of the observations in the sample space.

Example 2-1

A certain construction firm is designated to build a concrete bridge. It orders the concrete from a subcontractor who delivers one truckload (batch) of concrete to the site every quarter-hour. When making each batch of concrete, the subcontractor uses the same formula and types of ingredients.

A test cylinder of concrete is obtained from each batch of concrete. All cylinders are tested after 28 days, and the secant modulus of elasticity at that time is a random variable. If the measured concrete moduli from each cylinder is the same, then one would have reason to believe that the secant modulus of elasticity may be a deterministic quantity. However, such an observation is unlikely and, in general, one would expect to have different measurement values and thus the modulus would be a random variable. The numerical value of the modulus for each batch is called the *outcome*, or *observed value*, of the random variable. A collection of all possible outcomes of the random variable comprises its sample space. There are many possible events that can be associated with this random variable: for example, (1) the event that the measured modulus is greater than a prescribed numerical value, (2) the event that the measured modulus is greater than one prescribed value and less than another prescribed value.

In this example, the 28 day secant modulus was chosen as the random variable. However, this does not mean that it is the only random variable for this experiment. On the contrary, there are many other possible random variables that could be observed: for example, (1) the tangent modulus of elasticity, (2) the creep at 28 days, (3) the maximum compressive strength at 28 days, and (4) the maximum tensile strength at 28 days.

2-3 GRAPHIC DISPLAYS AND SAMPLE STATISTICS

The collection of data resulting from an experiment is only the first step in any study. The engineer must display the results of the test in a form that will be most meaningful to the intended application. With this goal in mind, both graphic and numerical measures are used by engineers to interpret the randomness of the results of an experiment.

A frequency *histogram* for one random variable is a graph that indicates the relative frequency that observed values of a random variable occur between two prescribed limits. Consider Example 2-1 of the previous section. It would be helpful to be able to answer a question such as: What percentage of the time is the secant modulus of elasticity between 3.0×10^6 psi and 3.3×10^6 psi? To illustrate how this question may be answered, imagine that in Example 2-1 the secant modulus was measured for 100 batches of concrete. Now, the 100 secant modulus of elasticity values are collected into 25 groups associated with the intervals shown in Table 2-1.

TABLE 2-1. Example: Concrete Cylinders

Modulus of Elasticity Interval ($\times 10^6$ psi)	*Number of Observations in Interval*	*Frequency of Occurrence*
Below 2.31	0	0.00
2.31–2.40	1	0.01
2.41–2.50	0	0.00
2.51–2.60	0	0.00
2.61–2.70	0	0.00
2.71–2.80	1	0.01
2.81–2.90	1	0.01
2.91–3.00	4	0.04
3.01–3.10	8	0.08
3.11–3.20	3	0.03
3.21–3.30	10	0.10
3.31–3.40	9	0.09
3.41–3.50	9	0.09
3.51–3.60	16	0.16
3.61–3.70	7	0.07
3.71–3.80	11	0.11
3.81–3.90	6	0.06
3.91–4.00	7	0.07
4.01–4.10	0	0.00
4.11–4.20	0	0.00
4.21–4.30	2	0.02
4.31–4.40	1	0.01
4.41–4.50	1	0.01
4.51–4.60	2	0.02
4.61–4.70	0	0.00
4.71–4.80	1	0.01
Above 4.80	0	0.00

When the number of observations in each increment is divided by 100 (the total number of observations), one obtains the relative frequency of occurrences of the random variable in each interval (see Table 2-1). These frequencies of occurrence are shown in bar graph form in Figure 2-1; such a graph is called a *frequency histogram*. The histogram shown in Figure 2-1 enables the engineer to state that, based on a sample size of 100, the modulus is between 3.0×10^6 psi and 4.0×10^6 psi, with a frequency of 0.86, or 86 percent of the time. Similarly, the modulus is less than 3.11×10^6 psi with a frequency of 0.15, or 15% of the time. Rather than saying that the frequency of occurrence is 0.15 or that the outcome occurs 15% of the time, an equivalent expression is that the *probability of the event* that the outcome is of the stated type is 0.15, or 15%.

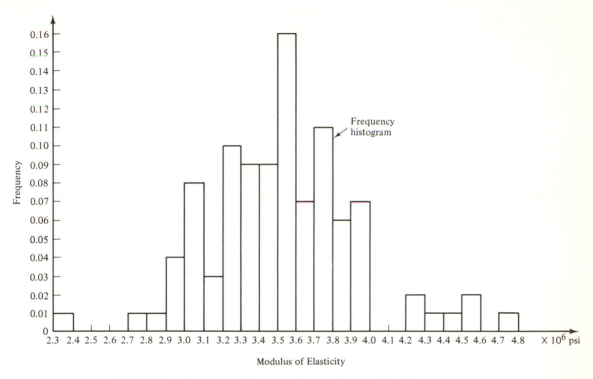

FIGURE 2-1. Frequency histogram for modulus of elasticity of concrete

A cumulative frequency histogram is a graph that indicates what frequency of observed values of a random variable are equal to or less than a prescribed value. The cumulative frequency histogram can be constructed from the original data or from a frequency histogram. Figure 2-2 shows the cumulative frequency histogram corresponding to the frequency histogram shown in Figure 2-1.

Two or more random variables may be observed simultaneously during a given experiment. The concept of a frequency histogram can be directly extended to multiple random variables. Consider Example 2-1, but now observe two random variables. One random variable is the secant modulus of elasticity (denoted by X_1), and the second is the maximum compressive strength of concrete (denoted by X_2). For each concrete batch (sample) the value of each random variable is observed and recorded. Table 2-2 shows the values of the first 20 batches of concrete.

A frequency histogram can now be constructed for each random variable (see Figure 2-3). In addition, a histogram for the two random variables can be created in the form of a three-dimensional bar graph. Each horizontal axis is divided into intervals which correspond to ranges of numerical outcomes of the random variable. Therefore, the horizontal plane is subdivided into rectangular areas. The vertical height of each rectangular bar is the frequency of occurrence, or probability, that for any one batch both random variables will have values in the intervals defined by

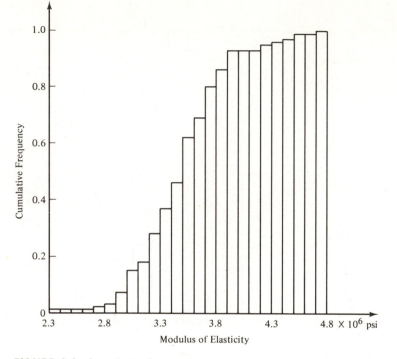

FIGURE 2-2. *Cumulative frequency histogram for modulus of concrete*

TABLE 2-2. Example: Two Random Variables

	Random Variables	
Sample Number	Secant Modulus (X_1) ($\times 10^6$ psi)	Compressive Strength (X_2) ($\times 10^3$ psi)
1	3.41	8.20
2	3.52	7.10
3	3.57	7.30
4	3.61	8.60
5	3.43	6.80
6	3.59	7.60
7	3.62	8.50
8	3.56	6.90
9	3.35	5.40
10	3.47	6.20
11	3.53	7.90
12	3.33	5.80
13	3.54	9.10
14	3.22	4.50
15	3.49	6.30
16	3.25	5.20
17	3.79	9.50
18	3.64	8.90
19	3.67	7.40
20	3.72	8.70

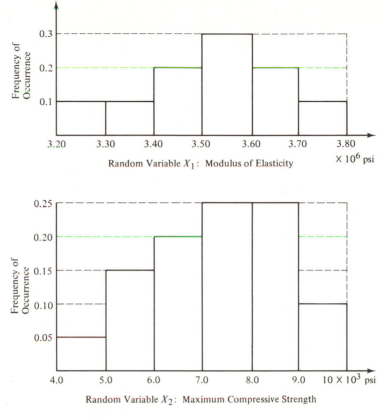

FIGURE 2-3. Example: Two histograms

the limits of the rectangular base. Figure 2-4 shows one rectangular region and the associated volume.

The frequency histograms of Figure 2-3 can be obtained from the three-dimensional frequency histogram shown in Figure 2-4 by summing the heights of all rectangular based volumes shown in Figure 2-4 for any interval. A cumulative frequency histogram for multiple random variables can be constructed in the same basic way as was done for one random variable. Figure 2-5 shows the cumulative frequency histogram corresponding to the frequency histogram shown in Figure 2-4.

Although a histogram graphically displays the observed data, it is usually desirable to have special numerical summaries which describe the data. The *sample mean* is a measure of central tendency of data and can be physically interpreted as the centroid of the data distribution. The sample mean corresponding to an observed set of data is defined as

$$ {}^s\bar{X}_1 \equiv \frac{1}{n} \sum_{j=1}^{n} {}^{(j)}x_1 \tag{2-1}$$

where the n outcomes of the experiment are denoted ${}^{(1)}x_1, {}^{(2)}x_1, \ldots, {}^{(n)}x_1$.

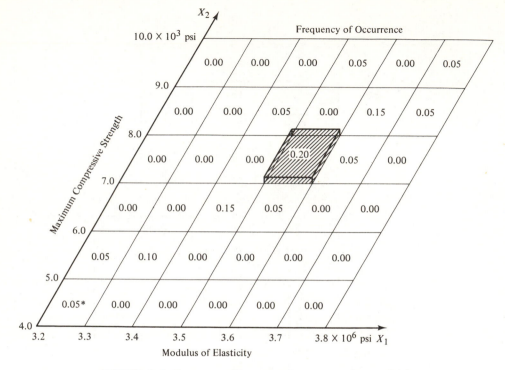

FIGURE 2-4. Frequency histogram for two random variables

The *sample variance* is a measure of the data's dispersion. Physically, the sample variance can be visualized as a statistical counterpart of area moment of inertia. The sample variance is defined as

$$^sX_1^2 \equiv \frac{1}{n} \sum_{j=1}^{n} [^{(j)}x_1 - {^s\bar{X}_1}]^2 \qquad (2\text{-}2)$$

There are sound reasons for using the denominator $(n-1)$ instead of n, and the reader is referred to reference [2.1]. However, here the more intuitively satisfying definition is used, and as the number of experiment observations increases, the numerical difference between values of the sample variances obtained using the two definitions decreases. The *sample standard deviation* is defined as the square root of the sample variance. The *sample coefficient of variation* is defined as the sample standard deviation divided by the sample mean. The *mth-order central statistical moment* is defined as

$$^sX_1^m \equiv \frac{1}{n} \sum_{j=1}^{n} [^{(j)}x_1 - {^s\bar{X}_1}]^m \qquad (2\text{-}3)$$

and the *mth-order statistical moment* (not central) is defined as

$$\hat{^s}X_1^m \equiv \frac{1}{n} \sum_{j=1}^{n} [^{(j)}x_1]^m \qquad (2\text{-}4)$$

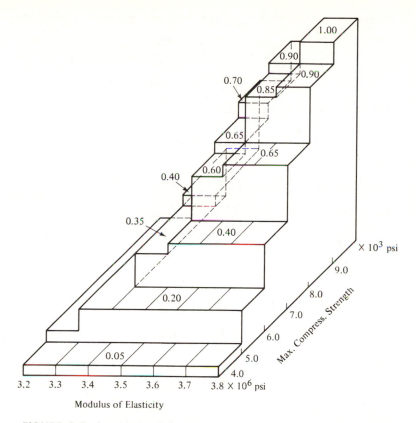

FIGURE 2-5. Cumulative frequency histogram for two random variables

If more than two random variables are observed, then the *sample covariance* between the random variables is defined as

$$^s\text{cov}(x_1,x_2) = \frac{1}{n}\sum_{j=1}^{n} [^{(j)}x_1 - {}^s\bar{X}_1][^{(j)}x_2 - {}^s\bar{X}_2] \qquad (2\text{-}5)$$

where the outcomes of the experiment are

$$^{(1)}x_1, {}^{(2)}x_1, \ldots, {}^{(n)}x_1 \quad \text{and} \quad {}^{(1)}x_2, {}^{(2)}x_2, \ldots, {}^{(n)}x_2$$

The *sample correlation coefficient* between two random variables is defined as the sample covariance divided by each random variable's standard deviation.

Example 2-2

Consider the experimental test data shown in Table 2-2. Calculate the sample mean, variance, standard deviation, and coefficient of variation for each random variable. Calculate the sample covariance and correlation coefficient relating the secant modulus of elasticity and the compressive strength.

Let

$$X_1 = \text{secant modulus of elasticity}$$
$$X_2 = \text{compressive strength}$$

The sample mean value of each random variable is

$$^s\bar{X}_1 \equiv \frac{1}{20} \sum_{j=1}^{20} {}^{(j)}x_1 = \frac{1}{20}(3.41 \times 10^6 + 3.52 \times 10^6 + \cdots) = 3.52 \times 10^6 \text{ psi}$$

$$^s\bar{X}_2 \equiv \frac{1}{20} \sum_{j=1}^{20} {}^{(j)}x_2 = \frac{1}{20}(8.20 \times 10^3 + 7.10 \times 10^3 + \cdots) = 7.30 \times 10^3 \text{ psi}$$

The sample variance of each random variable is

$$^s\bar{X}_1^2 \equiv \frac{1}{20} \sum_{j=1}^{20} [{}^{(j)}x_1 - {}^s\bar{X}_1]^2 = \frac{1}{20}[(3.41 \times 10^6 - 3.52 \times 10^6)^2 + \cdots]$$

$$= 2.13 \times 10^{10} \text{ (psi)}^2$$

$$^s\bar{X}_1^2 \equiv \frac{1}{20} \sum_{j=1}^{20} [{}^{(j)}x_2 - {}^s\bar{X}_2]^2 = \frac{1}{20}[(8.20 \times 10^3 - 7.30 \times 10^3)^2 + \cdots]$$

$$= 1.89 \times 10^6 \text{ (psi)}^2$$

The sample standard deviation of each random variable is:

$$\text{standard deviation of } X_1 = \sqrt{2.13 \times 10^{10}} = 0.15 \times 10^6 \text{ psi}$$

$$\text{standard deviation of } X_2 = \sqrt{1.89 \times 10^6} = 1.37 \times 10^3 \text{ psi}$$

The sample coefficient of variation of each random variable is

$$\text{coefficient of variation of } X_1 = \frac{0.15 \times 10^6}{3.52 \times 10^6} = 4.2\%$$

$$\text{coefficient of variation of } X_2 = \frac{1.37 \times 10^3}{7.30 \times 10^3} = 18.8\%$$

The sample covariance is

$$^s\text{cov}(x_1, x_2) = \frac{1}{20} \sum_{j=1}^{20} [{}^{(j)}x_1 - {}^s\bar{X}_1][{}^{(j)}x_2 - {}^s\bar{X}_2]$$

$$= \frac{1}{20}[(3.41 \times 10^6 - 3.52 \times 10^6)(8.20 \times 10^3 - 7.30 \times 10^3) + \cdots]$$

$$= 0.170 \times 10^9 \text{ (psi)}^2$$

The sample correlation coefficient is

$$\text{correlation coefficient} = \frac{0.170 \times 10^9}{(0.15 \times 10^6)(1.37 \times 10^3)} = 0.83$$

Example 2-3

A series of tests are conducted on A-36 structural steel. The sample mean and standard deviation of the yield stress are 47.9 ksi and 3.3 ksi, respectively. The sample mean and standard deviation of ultimate stress are 65.9 ksi and 2.9 ksi, respectively. Calculate the sample coefficient of variation for the yield stress and for the ultimate stress. Which stress has the greater uncertainty?

The sample coefficient of variations are

$$\text{yield stress} = \frac{3.3}{47.9} = 6.9\%$$

$$\text{ultimate stress} = \frac{2.9}{65.9} = 4.4\%$$

The greater uncertainty is in the yield stress (6.9 vs. 4.4),

Example 2-4

A series of tests are conducted to determine the statistical characteristics of the maximum 28 day compressive strength of good quality control concrete. The following sample statistics were measured:

$$\text{sample mean} = 5.19 \text{ ksi}$$
$$\text{sample standard deviation} = 550 \text{ psi}$$
$$\text{third central moment} = 1.23 \times 10^7 \text{ (psi)}^3$$
$$\text{fourth central moment} = 2.34 \times 10^{11} \text{ (psi)}^4$$

(a) Calculate the sample coefficient of variation.
(b) The *coefficient of skewness* is defined as the third central moment divided by the cube of the standard deviation. Calculate the coefficient of skewness.
(c) The *coefficient of kurtosis* is defined as the fourth central moment divided by the fourth power of the standard deviation. Calculate the coefficient of kurtosis.

The sample coefficient of variation is $(550/5190) = 10.6\%$. The coefficient of skewness is $(1.23 \times 10^7)/(550)^3 = 0.074$. The coefficient of kurtosis is $(2.34 \times 10^{11})/(550)^4 = 2.56$.

2-4 PROBABILITY DENSITY AND DISTRIBUTION FUNCTIONS

A histogram is very useful conceptually and quantitatively. However, it is usually desirable to have a continuous mathematical function that describes data. A *probability density function* (PDF) is such a mathematical function, and the degree to which it actually represents the information provided by a histogram is dependent upon its analytical complexity. For example, it can be seen from Figure 2-1 that a function must be selected having a very irregular shape in order to accurately represent the sampled data. This approach is usually not followed because the engineer works in reverse by selecting a probability density function that may be preferred for any one of several reasons (e.g., mathematical simplicity) and, at the same time, to a fair degree of accuracy, approximates the shape of the histogram. Values for the parameters of this PDF are then estimated by using the sampled data, as discussed in Section 2-7.

A probability density function can be visualized as follows. Assume that a random variable can take on any value on the real line. Now, for a segment along that line, e.g., from $\left(x - \frac{dx}{2}\right)$ to $\left(x + \frac{dx}{2}\right)$, the probability that the outcome of the experiment will fall in this interval is $p(x)\,dx$, where $p(x)$ is defined as the probability density function. Therefore, the probability that the random variable X will be between the two points on the real line, x_a and x_b, is

$$\Pr\,(x_a \le X \le x_b) = \int_{x_a}^{x_b} p(x)\,dx \tag{2-6}$$

The probability that the random variable will have a value equal to any set number on the real line is zero because $x_a = x_b$ in the above integral expression.

A probability density function *must* satisfy certain conditions. If a numerical-valued random variable is denoted by X and the corresponding probability density function by $p(x)$, and a lower-case x is used to denote an observed value of X, then the following conditions must be satisfied:

1. The probability density function is always equal to or greater than zero; i.e.,

$$p(x) \ge 0 \tag{2-7}$$

2. The area under the probability density function is unity; i.e.,

$$\int_{-\infty}^{+\infty} p(x)\,dx = 1 \tag{2-8}$$

3. The probability that an observed value of the random variable is between any two numerical values, e.g., x_a and x_b (denoted as $\Pr\,[x_a \le X \le x_b]$) is

$$\Pr[x_a \le X \le x_b] = \int_{x_a}^{x_b} p(x)\, dx \tag{2-9}$$

Any mathematical function that satisfies these three conditions can be chosen as a probability density function.

A *probability distribution function* is the integral of a probability density function. The probability distribution function is defined as

$$P(x_a) = \int_{-\infty}^{x_a} p(x)\, dx = \Pr[X \le x_a] \tag{2-10}$$

and $P(x_a)$ is the probability that the random variable X will have a value equal to or less than x_a.

If more than one random variable is associated with an experiment, then a *multivariate probability density function* is used to describe the random variables. Denote the random variables associated with a given experiment as $X_1, X_2, \ldots, X_n$. A multivariate probability density function must satisfy the following conditions:

1. $p(x_1, x_2, \ldots, x_n) \ge 0$ $\hspace{5cm}$ (2-11)

2. $\displaystyle\int_{-\infty}^{+\infty} \int_{-\infty}^{+\infty} \cdots \int_{-\infty}^{+\infty} p(x_1, x_2, \ldots, x_n)\, dx_1\, dx_2 \ldots dx_n = 1$ $\hspace{1cm}$ (2-12)

3. $\Pr[a_1 \le X_1 \le b_1, a_2 \le X_2 \le b_2, \ldots, a_n \le X_n \le b_n]$

$$= \int_{a_1}^{b_1} \int_{a_2}^{b_2} \cdots \int_{a_n}^{b_n} p(x_1, x_2, \ldots, x_n)\, dx_1\, dx_2 \ldots dx_n \tag{2-13}$$

A multiple random variable *probability distribution function* is defined by

$$P[x_a, x_b, \ldots, x_n] \equiv \int_{-\infty}^{x_a} \int_{-\infty}^{x_b} \cdots \int_{-\infty}^{x_n} p(x_1, x_2, \ldots, x_n)\, dx_1\, dx_2 \ldots dx_n$$

$$= \Pr[X_1 \le x_a, X_2 \le x_b, \ldots, X_n \le x_n] \tag{2-14}$$

where $x_a, x_b, \ldots, x_n$ are constants.

A probability density function for multiple random variables provides complete probabilistic information about the probability density function of any one random variable. A probability density function for one random variable obtained from a multivariate PDF is called a *marginal probability density function*. To illustrate, recall that the probability that X_j lies between a_j and b_j is given by

$$\Pr[a_j \le X_j \le b_j] = \Pr[a_j \le X_j \le b_j \text{ and } -\infty < X_k < \infty \text{ for all } k \ne j]$$

$$= \int_{x_j = a_j}^{x_j = b_j} \left[\int_{-\infty}^{+\infty} \int_{-\infty}^{+\infty} \cdots \int_{-\infty}^{+\infty} p(x_1, x_2, \ldots x_j, \ldots, x_n) \right.$$

$$\left. dx_1\, dx_2 \ldots dx_{j-1}\, dx_{j+1} \ldots dx_n \right] dx_j$$

$$= \int_{x_j = a_j}^{x_j = b_j} p(x_j)\, dx_j$$

where

$$p(x_j) = \int_{-\infty}^{+\infty} \int_{-\infty}^{+\infty} \cdots \int_{-\infty}^{+\infty} p(x_1, x_2, \ldots x_{j-1}, x_j, x_{j+1}, \ldots, x_n)$$

$$dx_1 \, dx_2 \ldots dx_{j-1} \, dx_{j+1} \ldots dx_n \qquad (2\text{-}15)$$

$p(x_j)$ is a marginal PDF.

Two random variables are said to be *independent* if their multivariate probability density function is equal to the product of the probability density function of each random variable; i.e.,

$$p(x_j, x_k) = p(x_j)p(x_k) \qquad (2\text{-}16)$$

Example 2-5

Consider the probability density function for extreme wind pressure, X, at a particular construction site to be

$$p(x) = \frac{C_0}{15} \qquad 5 \text{ psf} \leq X \leq 10 \text{ psf}$$

$$= 0 \qquad \text{otherwise}$$

What must the value of C_0 be such that the area under the PDF is unity? Calculate the probability distribution function for this random variable. What is the probability that the extreme wind pressure will be greater than 9 psf?

The area under the PDF must be unity, and therefore

$$\int_{-\infty}^{+\infty} p(x) \, dx = 1 = \int_5^{10} \frac{C_0}{15} \, dx = \frac{C_0}{3} \qquad \therefore \quad C_0 = 3$$

The probability distribution function is

$$P(x_a) = \int_{-\infty}^{x_a} p(x) \, dx$$

$$= 0 \qquad\qquad\qquad -\infty \leq x_a \leq 5 \text{ psf}$$

$$= \int_5^{x_a} \frac{1}{5} \, dx = \frac{x_a}{5} - 1 \qquad 5 \text{ psf} \leq x_a \leq 10 \text{ psf}$$

$$= 1 \qquad\qquad\qquad x_a > 10 \text{ psf}$$

The probability that the extreme wind pressure will be less than 9 psf is

$$P(9) = \frac{9}{5} - 1 = 0.80$$

Therefore, the probability that it will be greater than 9 psf is

$$\Pr[x_a > 9 \text{ psf}] = 1 - P(9) = 0.20$$

Example 2-6

If X_1 and X_2 are two random variables with a joint probability density function

$$p(x_1, x_2) = \left(\frac{1}{b-a}\right)\left(\frac{1}{d-c}\right) \qquad a \le X_1 \le b \text{ and } c \le X_2 \le d$$

$$= 0 \qquad\qquad\qquad \text{otherwise}$$

plot the probability density function for X_1 and X_2. Calculate the probability distribution function for X_1 and X_2. Calculate the marginal probability density function for X_1.

The PDF is

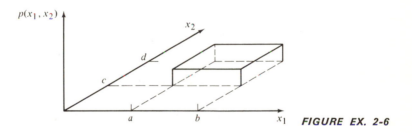

FIGURE EX. 2-6

The probability distribution function is

$$P(x_a, x_b) = \int_{-\infty}^{x_a} \int_{-\infty}^{x_b} p(x_1, x_2)\, dx_2\, dx_1$$

$$= 0 \qquad\qquad\qquad x_a < a, x_b < c$$

$$= \int_{a}^{x_a} \int_{c}^{x_b} \frac{1}{(b-a)(d-c)}\, dx_2\, dx_1$$

$$= \frac{(x_a - a)(x_b - c)}{(b-a)(d-c)} \qquad\qquad a \le x_a \le b, c \le x_b \le d$$

$$= \int_{-\infty}^{x_a} \int_{-\infty}^{x_b} \frac{1}{(b-c)(d-c)}\, dx_2\, dx_1$$

$$= \int_{-\infty}^{x_b} \left(\frac{1}{d-c}\right) dx_2 = \left(\frac{x_b - c}{d-c}\right) \qquad x_a > b, c \le x_b \le d$$

$$= \int_{-\infty}^{x_a} \int_{-\infty}^{x_b} \frac{1}{(b-a)(d-c)} \, dx_2 \, dx_1$$

$$= \int_{-\infty}^{x_a} \left(\frac{1}{b-a}\right) dx_1 = \left(\frac{x_a - a}{b-a}\right) \qquad a \le x_a \le b, \, x_b > d$$

$$= 1 \qquad\qquad\qquad\qquad x_a > b, \, x_b > d$$

The marginal PDF for X_1 is

$$p(x_1) = \int_{-\infty}^{+\infty} p(x_1, x_2) \, dx_2$$

$$= \int_{c}^{d} \frac{1}{(b-a)(d-c)} \, dx_2 = \left(\frac{1}{b-a}\right) \qquad a \le X_1 \le b,$$

$$= 0 \qquad\qquad\qquad\qquad\qquad\qquad \text{otherwise}$$

This multivariate probability density function is given a special name because of its wide usage. It is called a *two-dimensional* (or *bivariate*) *UNIFORM probability density function*.

2-5 EXPECTATION OF RANDOM VARIABLES

The mathematical operation of expectation is a weighted integration of a random variable. The expectation of a function of a random variable is defined as

$$E\langle g(X) \rangle \equiv \int_{-\infty}^{+\infty} g(x) p(x) \, dx \qquad\qquad (2\text{-}17)$$

where $g(x)$ is the function of the random variable. The operation of expectation can be visualized by using Fig. 2-6. This figure shows that the expectation of a function of a random variable is equal to the area under the curve defined by $g(x)$ times $p(x)$.

The most fundamental expectation is the *mean*, or average, value of a random variable, and it is defined as

$$E\langle X \rangle \equiv \int_{-\infty}^{+\infty} x p(x) \, dx \qquad\qquad (2\text{-}18)$$

The mean is a constant and will hereafter be denoted with a bar over the top of the random variable, i.e., $\bar{X}$. This operation sums, for all possible values of the random variable, the product of the random variable and the probability of its occurrence.

Special and useful examples of the expectation of a function of a random variable are as follows:

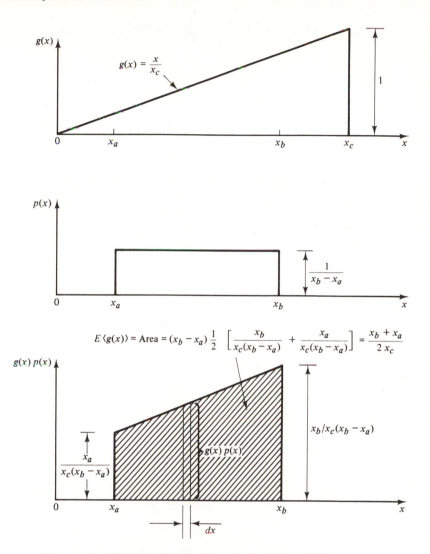

FIGURE 2-6. *Expectation of a random variable*

1. The expected value of a constant, noted C, times the random variable is

$$E\langle CX \rangle = C\bar{X} \qquad (2\text{-}19)$$

2. The *mean square value* of a random variable is defined as

$$E\langle X^2 \rangle \equiv \int_{-\infty}^{+\infty} x^2 p(x)\, dx \qquad (2\text{-}20)$$

3. The *variance* of a random variable is denoted Var (X) or σ_X^2 and is defined as

$$\text{Var}\,(X) \equiv E\langle(X - \bar{X})^2\rangle = \int_{-\infty}^{+\infty} (x - \bar{X})^2 p(x)\, dx$$
$$= E\langle X^2\rangle - (\bar{X})^2 \tag{2-21}$$

4. The *standard deviation* of a random variable is defined as

$$\sigma_x = +\sqrt{\text{Var}\,(X)} \tag{2-22}$$

5. The *mth central moment* of a random variable is defined as

$$E\langle(X - \bar{X})^m\rangle \equiv \int_{-\infty}^{+\infty} (x - \bar{X})^m p(x)\, dx \tag{2-23}$$

The *coefficient of variation* of a random variable is its standard deviation divided by its mean.

Expectation also applies to a set of random variables. If there are n random variables, denoted $X_1, X_2, \ldots, X_n$, and their corresponding probability density function is $p(x_1, x_2, \ldots, x_n)$, then the expectation of a general function of all random variables $g(x_1, x_2, \ldots, x_n)$ is defined as

$$E\langle g(X_1, X_2, \ldots, X_n)\rangle$$
$$= \int_{x_1=-\infty}^{+\infty} \cdots \int_{x_n=-\infty}^{+\infty} g(x_1, x_2, \ldots, x_n) p(x_1, x_2, \ldots x_n)\, dx_1\, dx_2 \ldots dx_n \tag{2-24}$$

Certain expectations are used often enough to warrant special names. These expectations are:

1. The *cross-correlation* between X_j and X_k is defined as

$$E\langle X_j X_k\rangle = \int_{x_j=-\infty}^{+\infty} \int_{x_k=-\infty}^{+\infty} x_j x_k\, p(x_j, x_k)\, dx_j\, dx_k \tag{2-25}$$

where $p(x_j, x_k)$ is the joint PDF for X_j, and X_k and is given by

$$p(x_j, x_k) = \int_{x_1=-\infty}^{+\infty} \cdots \int_{x_{j-1}=-\infty}^{+\infty} \int_{x_{j+1}=-\infty}^{+\infty} \cdots \int_{x_{k-1}=-\infty}^{+\infty} \int_{x_{k+1}=-\infty}^{+\infty} \cdots \int_{x_n=-\infty}^{+\infty}$$
$$p(x_1, x_2, \ldots, x_n)\, dx_1\, dx_2 \ldots dx_{j-1}\, dx_{j+1} \ldots dx_{k-1}$$
$$dx_{k+1} \ldots dx_n \tag{2-26}$$

2. The *covariance* between X_j and X_k is defined as

$$\text{Cov}\,(X_j, X_k) \equiv E\langle(X_j - \bar{X}_j)(X_k - \bar{X}_k)\rangle$$
$$= \int_{x_j=-\infty}^{+\infty} \int_{x_k=-\infty}^{+\infty} (x_j - \bar{X}_j)(x_k - \bar{X}_k) p(x_j, x_k)\, dx_j\, dx_k \tag{2-27}$$

where $\bar{X}_j$ and $\bar{X}_k$ are the mean values of X_j and X_k.

3. A nondimensional term that is useful is the *correlation coefficient,* which is defined as

$$\rho_{jk} = \frac{\text{Cov } (X_j, X_k)}{\sqrt{\text{Var}(X_j) \text{ Var } (X_k)}} \tag{2-28}$$

Recall that two random variables X_j and X_k are said to be *independent* if and only if

$$p(x_j, x_k) = p(x_j)p(x_k) \tag{2-29}$$

If X_j and X_k are independent, then it follows from Eq. (2-27) that

$$\text{Cov } (X_j, X_k) = 0 \tag{2-30}$$

Two independent random variables have a zero covariance. However, a covariance of zero can exist without the random variables being independent.

The *mean vector* of a set of n random variables, $X_1, X_2, \ldots, X_n$, is defined as

$$\{\bar{X}\} = \begin{Bmatrix} \bar{X}_1 \\ \bar{X}_2 \\ \cdot \\ \cdot \\ \cdot \\ X_n \end{Bmatrix}$$

and their *covariance matrix* is defined as

$$[S_X] = \begin{bmatrix} \text{Var } (X_1) & \text{Cov } (X_1, X_2) & \cdots & \text{Cov } (X_1, X_n) \\ \text{Cov } (X_2, X_1) & \text{Var } (X_2) & \cdots & \text{Cov } (X_2, X_n) \\ \cdot & & & \cdot \\ \cdot & & & \cdot \\ \text{Cov } (X_n, X_1) & \text{Cov } (X_1, X_2) & \cdots & \text{Var } (X_n) \end{bmatrix}$$

Example 2-7

(a) The deflection at the end of a cantilever beam due to a concentrated load at its free end is

$$\Delta = \frac{Pl^3}{3EI}$$

where

$$P = \text{loading}$$
$$l = \text{length of beam}$$
$$E = \text{modulus of elasticity}$$
$$I = \text{moment of inertia}$$

If l, E, and I are deterministic and P is a random variable with mean $\bar{P}$ and standard deviation σ_P, then calculate the mean and standard deviation of Δ.

(b) The support moment for a fixed-fixed beam acted upon by a random concentrated load at midspan and a random uniformly distributed load is

$$M = \frac{Pl}{8} + \frac{\omega l^2}{12}$$

where

$$P = \text{concentrated load at midspan}$$

$$\omega = \text{uniformly distributed load}$$

$$l = \text{length of the beam}$$

Let l be deterministic. If the means and standard deviations of the loads are $\bar{P}$, $\bar{\omega}$, σ_P, σ_ω and $\text{Cov}\,(P, \omega) = 0$, then find the mean and standard deviation of the support moment.

(c) Consider the same problem as described in (b), but in addition consider the midspan deflection given by

$$\Delta = \frac{Pl^3}{192EI} + \frac{\omega l^4}{384EI}$$

where

$$E = \text{modulus of elasticity}$$

$$I = \text{moment of inertia}$$

Let E and I be deterministic. Calculate the mean and standard deviation of Δ plus the covariance relating M and Δ.

Part (a) solution:

$$\bar{\Delta} = \text{mean of } \Delta = E\langle\Delta\rangle = E\left\langle\frac{Pl^3}{3EI}\right\rangle = \left(\frac{l^3}{3EI}\right)E\langle P\rangle = \frac{\bar{P}l^3}{3EI}$$

$$\sigma_\Delta^2 = \text{Var}\,(\Delta) = E\langle(\Delta - \bar{\Delta})^2\rangle = E\left\langle\left(\frac{Pl^3}{3EI} - \frac{\bar{P}l^3}{3EI}\right)^2\right\rangle$$

$$= \left(\frac{l^3}{3EI}\right)^2 E\langle(P - \bar{P})^2\rangle = \left(\frac{l^3}{3EI}\right)^2 \sigma_P^2$$

$$\sigma_\Delta = \text{standard deviation of } \Delta = \left(\frac{l^3}{3EI}\right)\sigma_P$$

Part (b) solution:

$$\bar{M} = \text{mean of } M = E\langle M\rangle = E\left\langle\frac{Pl}{8} + \frac{\omega l^2}{12}\right\rangle = \frac{\bar{P}l}{8} + \frac{\bar{\omega}l^2}{12}$$

$$\sigma_M^2 = \text{Var}(M) = E\langle(M - \bar{M})^2\rangle = E\left\langle\left(\frac{Pl}{8} + \frac{\omega l^2}{12} - \frac{\bar{P}l}{8} - \frac{\bar{\omega}l^2}{12}\right)^2\right\rangle$$

$$= E\left\langle\left[\frac{l}{8}(P - \bar{P}) + \frac{l^2}{12}(\omega - \bar{\omega})\right]^2\right\rangle$$

$$= \left(\frac{l}{8}\right)^2\sigma_P^2 + \left(\frac{l^2}{12}\right)^2\sigma_\omega^2 + 2\left(\frac{l}{8}\right)\left(\frac{l^2}{12}\right)\underbrace{\text{Cov}(P, \omega)}_{0}$$

$$\sigma_M = \text{standard deviation of } M = \sqrt{\left(\frac{l}{8}\right)^2\sigma_P^2 + \left(\frac{l^2}{12}\right)^2\sigma_\omega^2}$$

Note that the variances of P and ω are in effect added to obtain the variance of M.

Part (c) solution:

$$\bar{\Delta} = \text{mean of } \Delta = \frac{\bar{P}l^3}{192EI} + \frac{\bar{\omega}l^4}{384EI}$$

$$\sigma_\Delta = \text{standard deviation of } \Delta = \sqrt{\left(\frac{l^3}{192EI}\right)^2\sigma_P^2 + \left(\frac{l^4}{384EI}\right)^2\sigma_\omega^2}$$

$$\text{Cov}(M, \Delta) = \text{covariance of } M \text{ and } \Delta$$

$$= E\langle(M - \bar{M})(\Delta - \bar{\Delta})\rangle$$

$$= E\left\langle\left[\frac{l}{8}(P - \bar{P}) + \frac{l^2}{12}(\omega - \bar{\omega})\right]\left[\frac{l^3}{192EI}(P - \bar{P}) + \frac{l^4}{384EI}(\omega - \bar{\omega})\right]\right\rangle$$

$$= \left(\frac{l^4}{1536}\right)\sigma_P^2 + \left(\frac{l^6}{4608EI}\right)\sigma_\omega^2$$

Example 2-8

Consider the probability density function given in Example 2-5; i.e.,

$$p(x) = \frac{1}{5} \qquad 5 \text{ psf} \leq X \leq 10 \text{ psf}$$

$$= 0 \qquad \text{otherwise}$$

Calculate the mean, variance, standard deviation, and coefficient of variation of X.

$$\bar{X} = \text{mean} = \int_{-\infty}^{+\infty} xp(x)\, dx = \int_5^{10} x\left(\frac{1}{5}\right) dx = 7.5 \text{ psf}$$

$$\sigma_X^2 = \text{variance} = \int_{-\infty}^{+\infty} (x - \bar{X})^2 p(x)\, dx = \int_5^{10} (x - 7.5)^2\left(\frac{1}{5}\right) dx = 2.08 \text{ (psf)}^2$$

$$\sigma_x = \text{standard deviation} = \sqrt{2.08} = 1.44 \text{ psf}$$

$$\rho_x = \text{coefficient of variation} = \frac{\sigma_x}{\bar{X}} = \frac{1.44}{7.50} = 0.19 \ (19\%)$$

Example 2-9

Define

$$Y \equiv X_1 + X_2$$

where both X_1 and X_2 are random variables. Let

$$\bar{X}_1, \bar{X}_2 = \text{mean of } X_1 \text{ and } X_2, \text{ respectively}$$

$$\sigma_{X_1}, \sigma_{X_2} = \text{standard deviation of } X_1 \text{ and } X_2, \text{ respectively}$$

$$\rho_{X_1, X_2} = \text{correlation coefficient between } X_1 \text{ and } X_2$$

$$= \frac{\text{Cov}\,(X_1,\,X_2)}{\sigma_{X_1}\sigma_{X_2}}$$

$$\text{Cov}\,(X_1,\,X_2) = \text{covariance of } X_1 \text{ and } X_2$$

Calculate the mean and standard deviation of Y.

The mean of Y is

$$E\langle Y \rangle = E\langle (X_1 + X_2) \rangle = E\langle X_1 \rangle + E\langle X_2 \rangle = \bar{X}_1 + \bar{X}_2$$

Therefore, the mean is the sum of the means of the random variables which are added. The variance of Y is

$$\begin{aligned}
E\langle (Y - E\langle Y \rangle)^2 \rangle &= E\langle (X_1 + X_2 - \bar{X}_1 - \bar{X}_2)^2 \rangle \\
&= E\langle [(X_1 - \bar{X}_1) + (X_2 - \bar{X}_2)]^2 \rangle \\
&= E\langle (X_1 - \bar{X}_1)^2 + (X_2 - \bar{X}_2)^2 + 2(X_1 - \bar{X}_1)(X_2 - \bar{X}_2) \rangle \\
&= \sigma_{X_1}^2 + \sigma_{X_2}^2 + 2\,\text{Cov}\,(X_1,\,X_2)
\end{aligned}$$

Example 2-10

Consider the PDF given in Example 2-6. Calculate the mean of X_1 and the covariance of X_1 and X_2.

In Example 2-6, the marginal PDF for X_1 was shown to be

$$p(x_1) = \left(\frac{1}{b-a}\right) \qquad a \le X_1 \le b$$

$$= 0 \qquad \text{otherwise}$$

Therefore, the mean of X_1 is

$$\bar{X}_1 = E\langle X_1 \rangle = \int_{-\infty}^{+\infty} x_1 p(x_1)\, dx_1 = \int_a^b x_1\left(\frac{1}{b-a}\right) dx_1 = \frac{a+b}{2}$$

The covariance of X_1 and X_2 is

$$\text{Cov}\,(X_1, X_2) = \int_{-\infty}^{+\infty}\int_{-\infty}^{+\infty} (x_1 - \bar{X}_1)(x_2 - \bar{X}_2) p(x_1, x_2)\, dx_1\, dx_2$$

It was shown that

$$\bar{X}_1 = \left(\frac{a+b}{2}\right)$$

and similarly it follows that

$$\bar{X}_2 = \left(\frac{c+d}{2}\right)$$

Thus,

$$\text{Cov}\,(X_1, X_2) = \int_a^b \int_c^d (x_1 - \bar{X}_1)(x_2 - \bar{X}_2)\left[\frac{1}{(b-a)(d-c)}\right] dx_2\, dx_1$$

$$= \int_a^b \left[\frac{1}{(b-a)(d-c)}\right](x_1 - \bar{X}_1)\left\{\underbrace{\int_a^d (x_2 - \bar{X}_2)\, dx_2}_{0}\right\} dx_1$$

$$= 0$$

2-6 COMMON PROBABILITY DENSITY FUNCTIONS

Section 2-4 specifies three conditions which a mathematical function must satisfy to be a probability density function. There are several probability density functions which are used frequently in structural engineering and which will be used in this book. In this section we define these probability density functions and some of their properties.

Uniform (Two-Parameter Function: a and b)

The uniform PDF is defined as

$$p(x) = \left(\frac{1}{b-a}\right) \qquad a \leq x \leq b$$
$$= 0 \qquad\qquad \text{otherwise}$$

(2-31)

and Figure 2-7 shows its general shape.

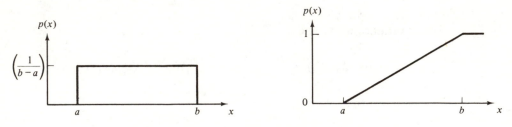

FIGURE 2-7. Uniform probability density function.

The mean of X is

$$\bar{X} = \int_{-\infty}^{+\infty} xp(x)\,dx = \left(\frac{a+b}{2}\right)$$

(2-32)

and its variance is

$$\sigma_X^2 = \int_{-\infty}^{+\infty} (x - \bar{X})^2 p(x)\,dx = \frac{(b-a)^2}{12}$$

(2-33)

Normal (Two-Parameter Function: a and b)

The normal PDF is shown in Figure 2-8 and is defined as

$$p(x) = \frac{1}{a\sqrt{2\pi}} \exp\left\{-\frac{1}{2}\left(\frac{x-b}{a}\right)^2\right\} \qquad \begin{array}{l} -\infty \leq x \leq +\infty \\ a > 0,\, b > -\infty \end{array}$$

(2-34)

The mean of X is

$$\bar{X} = b$$

(2-35)

and its variance is

$$\sigma_X^2 = a^2$$

(2-36)

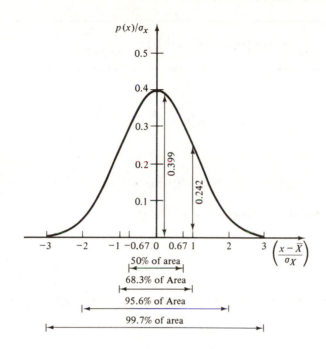

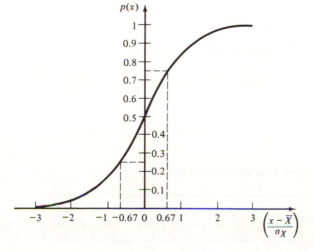

FIGURE 2-8. Normal probability density and distribution functions

27

Log-Normal (Two-Parameter Function: a and b)

$$p(x) = \frac{1}{ax\sqrt{2\pi}} \exp\left\{-\frac{1}{2}\left(\frac{\ln x - b}{a}\right)^2\right\} \qquad \begin{array}{l} 0 \leq x \leq +\infty \\ a > 0, b > -\infty \end{array} \qquad \text{(2-37)}$$

The mean of X is

$$\bar{X} = \exp\left\{b + \frac{a^2}{2}\right\} \qquad \text{(2-38)}$$

and its variance is

$$\sigma_X^2 = (\exp\{2b + a^2\})(\exp\{a^2\} - 1) \qquad \text{(2-39)}$$

Log-Normal (Three-Parameter Function: a, b, and c)

$$p(x) = \frac{1}{(x - c)a\sqrt{2\pi}} \exp\left\{-\frac{1}{2}\left(\frac{\ln(x - c) - b}{a}\right)^2\right\} \qquad \begin{array}{l} c \leq x \leq +\infty \\ a > 0, b < \infty \\ c > -\infty \end{array} \qquad \text{(2-40)}$$

The mean of X is

$$\bar{X} = c + \exp\left\{b + \frac{a^2}{2}\right\} \qquad \text{(2-41)}$$

and its variance is

$$\sigma_X^2 = (\exp\{2b + a^2\})(\exp\{a^2\} - 1) \qquad \text{(2-42)}$$

The third moment of X about the mean is

$$(\bar{X})^3(\rho^6 + 3\rho^4) \qquad \text{(2-43)}$$

where

$$\rho = \text{coefficient of variation} = \frac{\sigma_X}{\bar{X}}$$

Gamma (Two-Parameter Function: a and b)

$$p(x) = \frac{1}{a\Gamma(b)}\left(\frac{x}{a}\right)^{b-1} \exp\left\{-\frac{x}{a}\right\} \qquad \begin{array}{l} x > 0 \\ a, b > 0 \end{array} \qquad \text{(2-44)}$$

where $\Gamma(\ )$ is the gamma function

$$\Gamma(b) = \int_0^\infty y^{b-1} \exp\{-y\}\, dy \qquad \text{(2-45)}$$

which is equal to $(b - 1)!$ when b is a positive integer. For noninteger values of b the gamma function can be found in standard mathematical handbooks.
The mean of X is

$$\bar{X} = ab \tag{2-46}$$

and the variance is

$$\sigma_X^2 = a^2 b \tag{2-47}$$

when the parameter b becomes large the shape of the central portion of the gamma PDF is well approximated by a normal PDF. Using the variable transformation

$$Z \equiv \frac{X}{a}$$

it follows that Eq. (2-44) becomes

$$p(z) = \frac{1}{\Gamma(b)} z^{b-1} \exp\{-z\}, \qquad 0 < z \tag{2-48}$$

Figure 2-9 shows the form of Eq. (2-48) for several values of b. When b is less than unity the PDF has a singularity at $z = 0$.

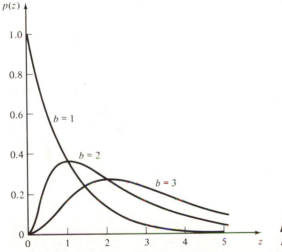

FIGURE 2-9. Gamma probability density function

Gamma (Three-Parameter Function: a, b, and c)

$$p(x) = \frac{1}{a\Gamma(b)} \left(\frac{x - c}{a}\right)^{b-1} \exp\left\{-\left(\frac{x - c}{a}\right)\right\} \qquad \begin{array}{c} x > c \\ a, b > 0 \end{array} \tag{2-49}$$

where the mean of X is

$$\bar{X} = c + ab \tag{2-50}$$

and the variance is

$$\sigma_X^2 = a^2 b \tag{2-51}$$

Exponential (Two-Parameter Function: a and c)

The exponential PDF is a special case of the three-parameter gamma PDF. If $b = 1$, then

$$p(x) = \frac{1}{a} \exp\left\{-\left(\frac{x-c}{a}\right)\right\} \qquad \begin{array}{l} x > c \\ a > 0 \end{array} \tag{2-52}$$

Beta (Four-Parameter Function: a, b, c, and d)

$$p(x) = \left(\frac{1}{b-a}\right)[B(a,b)]^{-1}\left(\frac{x-a}{b-a}\right)^{c-1}\left[1-\left(\frac{x-a}{b-a}\right)\right]^{d-1} \qquad \begin{array}{l} a \le x \le b \\ 0 \le a < b \\ c, d > 0 \end{array} \tag{2-53}$$

where

$$B(a,b) \equiv \frac{\Gamma(a)\Gamma(b)}{\Gamma(a+b)} \tag{2-54}$$

The mean of X is

$$\bar{X} = a + (b-a)\left(\frac{c}{c+d}\right) \tag{2-55}$$

and the variance of X is

$$\sigma_X^2 = (b-a)^2\left[\frac{cd}{(c+d)^2(c+d+1)}\right] \tag{2-56}$$

It also follows that

$$E\langle(X-\bar{X})^3\rangle = (b-a)^3\left[\frac{2cd(d-c)}{(c+d)^3(c+d+1)(c+d+2)}\right] \tag{2-57}$$

and

$$E\langle(X-\bar{X})^4\rangle = (b-a)^4\left\{\frac{3cd[2(c-d)^2 + cd(c+d+2)]}{(c+d)^4(c+d+1)(c+d+2)(c+d+3)}\right\} \tag{2-58}$$

The uniform PDF is special case of the beta PDF if $c = 0$ and $d = 1$.

Type I Extreme Value (Two-Parameter Function: a and b)

The type I extreme value PDF is also called a *Fisher-Tippett type I* or a *Gumbel*, and assumes two forms. The form usually used for the *largest extreme value* is

$$p(x) = a\exp\left[-a(x-b) - e^{-a(x-b)}\right] \qquad \begin{array}{l} -\infty \le x \le +\infty \\ a > 0, b < \infty \end{array} \tag{2-59}$$

The mean of X is

$$\bar{X} = b + \left(\frac{\gamma}{a}\right) \tag{2-60}$$

where γ is the Euler constant and is an infinite sequence of numbers with the first three numbers 0.577, and therefore

$$\bar{X} \cong b + \left(\frac{0.577}{a}\right)$$

The variance of X is

$$\sigma_X^2 = \frac{\pi^2}{6a^2} \tag{2-61}$$

The corresponding PDF used for the *smallest extreme value* is

$$\boxed{p(x) = a \exp\left[a(x-b) - e^{a(x-b)}\right] \qquad \begin{array}{l} -\infty \le x \le +\infty \\ a > 0, b < \infty \end{array}} \tag{2-62}$$

and its distribution function is

$$P(x) = 1 - \exp\left[-e^{a(x-b)}\right] \qquad -\infty \le x \le +\infty$$

with

$$\bar{X} = \text{mean} = b - \frac{\gamma}{a} \tag{2-63}$$

and

$$\sigma_X^2 = \text{variance} = \frac{\pi^2}{6a^2} \tag{2-64}$$

Type II Extreme Value (Two-Parameter Function: a and b)

The PDF of the type II extreme value distribution is

$$\boxed{p(x) = \left(\frac{a}{b}\right)\left(\frac{b}{x}\right)^{a+1} \exp\left\{-\left(\frac{b}{x}\right)^a\right\} \qquad \begin{array}{l} x \ge 0 \\ a, b > 0 \end{array}} \tag{2-65}$$

This density function is also referred to as a *Fisher-Tippett type II* or a *Fréchet*. The mean of X is

$$\bar{X} = b\Gamma\left[1 - \left(\frac{1}{a}\right)\right] \tag{2-66}$$

and its variance is

$$\sigma_X^2 = b^2\left\{\Gamma\left[1 - \left(\frac{2}{a}\right)\right] - \Gamma^2\left[1 - \left(\frac{1}{a}\right)\right]\right\} \tag{2-67}$$

Weibull (Three-Parameter Function: a, b, and c)

The PDF of the Weibull is

$$p(x) = \left(\frac{a}{b}\right)\left(\frac{x-c}{b}\right)^{a-1} \exp\left\{-\left(\frac{x-c}{b}\right)^a\right\} \qquad \begin{array}{l} c < x < +\infty \\ a, b > 0, c > -\infty \end{array} \qquad (2\text{-}68)$$

The mean of X is

$$\bar{X} = c + b\Gamma\left[1 + \left(\frac{1}{a}\right)\right] \qquad (2\text{-}69)$$

and its variance is

$$\sigma_x^2 = b^2\left\{\Gamma\left[1 + \left(\frac{2}{a}\right)\right] - \Gamma^2\left[1 + \left(\frac{1}{a}\right)\right]\right\} \qquad (2\text{-}70)$$

The Weibull is a special case of the largest type I extreme value distribution where the transformation

$$X = \ln\left(\frac{1}{\tilde{X}}\right)$$

has been used, and the PDF of $\tilde{X}$ is Eq. (2-59) and the PDF of X is Eq. (2-68).

If $a = 1$ and $c = 0$ in Eq. (2-68), then the PDF is identical with an exponential. The Rayleigh PDF is also a special case of the Weibull, with $a = 2$, $c = 0$, and the parameter b replaced by a

$$b^* \equiv \sqrt{2}\, b$$

Only the uniform and normal multivariate probability density functions are used in this book. Prior to considering the most general case, consider the following cases where only two random variables exist.

The *bivariate or joint uniform PDF* is defined as

$$\begin{aligned} p(x_1, x_2) &= \frac{1}{(b-a)(d-c)} \qquad a \le x_1 \le b, c \le x_2 \le d \\ &= 0 \qquad\qquad\qquad \text{otherwise} \end{aligned} \qquad (2\text{-}71)$$

This PDF is discussed in Examples 2-6 and 2-10.

The *bivariate or joint normal PDF* is defined as

$$\begin{aligned} p(x_1, x_2) &= \frac{1}{2\pi\sigma_1\sigma_2\sqrt{1-\rho^2}} \\ &\exp\left\{\frac{-1}{2(1-\rho^2)}\left[\left(\frac{x_1-m_1}{\sigma_1}\right)^2 + \left(\frac{x_2-m_2}{\sigma_2}\right)^2 - 2\rho\left(\frac{x_1-m_1}{\sigma_1}\right)\left(\frac{x_2-m_2}{\sigma_2}\right)\right]\right\} \end{aligned} \qquad (2\text{-}72)$$

where

$$m_1 = \bar{X}_1$$
$$m_2 = \bar{X}_2$$
$$\sigma_1^2 = \text{Var}(X_1)$$
$$\sigma_2^2 = \text{Var}(X_2)$$
$$\rho = \frac{\text{Cov}(X_1, X_2)}{\sigma_1 \sigma_2}$$

The general *multivariate uniform PDF* is

$$p(x_1, x_2, \ldots, x_n) = \frac{1}{(b_1 - a_1)(b_2 - a_2)\ldots(b_n - a_n)} \qquad a_j \leq x_j \leq b_j$$
$$= 0 \qquad\qquad\qquad\qquad\qquad \text{otherwise}$$

(2-73)

The general *multivariate normal PDF* is

$$p(x_1, x_2, \ldots, x_n)$$
$$\equiv \frac{1}{(2\pi)^{n/2}} \frac{1}{|S_x|^{1/2}} \exp\left\{ -\frac{1}{2} \sum_{j,k=1}^{n} (x_j - \bar{x}_j) S_x^{jk} (x_k - \bar{x}_k) \right\}$$

(2-74)

with the *covariance matrix* and its inverse defined as

$$[S_x] = \begin{bmatrix} S_{11}^x & S_{12}^x & \cdots & S_{1n}^x \\ S_{21}^x & S_{22}^x & \cdots & S_{2n}^x \\ \cdot & & & \\ \cdot & & & \\ \cdot & & & \\ S_{n1}^x & S_{n2}^x & \cdots & S_{nn}^x \end{bmatrix} = \begin{bmatrix} \text{Var}(X_1) & \text{Cov}(X_1, X_2) & \cdots & \text{Cov}(X_1, X_n) \\ \text{Cov}(X_2, X_1) & \text{Var}(X_2) & \cdots & \text{Cov}(X_2, X_n) \\ \cdot & & & \\ \cdot & & & \\ \text{Cov}(X_n, X_1) & \text{Cov}(X_n, X_1) & \cdots & \text{Var}(X_n) \end{bmatrix}$$

and

$$[S_x]^{-1} = \begin{bmatrix} S_x^{11} & S_x^{12} & \cdots & S_x^{1n} \\ S_x^{21} & S_x^{22} & & S_x^{2n} \\ \cdot & & & \\ \cdot & & & \\ \cdot & & & \\ S_x^{n1} & S_x^{n2} & \cdots & S_x^{nn} \end{bmatrix}$$

respectively.

Also $|S_x|^{1/2}$ denotes the determinant of the covariance matrix, $[S_x]$, raised to the one-half power.

2-7 SAMPLE STATISTICS TO PROBABILITY DENSITY FUNCTIONS

In Section 2-3 we discussed the calculation of sample statistics from data, and in Sections 2-4 through 2-6 we reviewed the mathematical development of *probability theory*.

To place these sections in proper perspective, it should be recognized that an important area of statistics is concerned with estimating the parameters of a probabilistic model (i.e., a PDF). Herein, the mean, variance(s), covariances, and any moments of a probabilistic model are estimated by using the sample statistics of the corresponding parameters defined in Section 2-3. That is, the sample mean is set equal to the mean of the PDF, etc. Other techniques exist for estimating the parameters of a PDF; however, the above is simple and usually provides estimates which are sufficiently accurate for structural engineering applications. The reader is referred to references [2.1] and [2.2].

Example 2-11

A series of tests are conducted on reinforced concrete beams without web reinforcement. The random variable, denoted R, is

$$R = \left(\frac{\text{test shear strength}}{\text{design formula shear strength}}\right)$$

where

V_c = design formula shear strength

$$= (bd)\left[1.9\sqrt{f'_c} + 2500\rho\left(\frac{Vd}{M}\right)\right] \leq 3.5\sqrt{f'_c}$$

f'_c = maximum 28 day compressive stress of concrete

ρ = percentage of reinforcing steel

b = beam width

d = effective depth

M = bending moment

V = shear force

The sample mean and standard deviation of R are 1.094 and 0.167, respectively. Calculate the sample coefficient of variation. Assume that R has a normal PDF, and define the PDF.

Sample coefficient of variation $= (\sigma_R/\bar{R}) = 0.167/1.094 = 15.3\%$.

$$p(r) = \frac{1}{\sqrt{2\pi}\,(0.167)} \exp\left[-\frac{1}{2}\left(\frac{r - 1.094}{0.167}\right)^2\right]$$

Example 2-12

A series of tests are performed on square tied reinforced concrete columns. The random variable, denoted R, is

$$R = \left(\frac{\text{test ultimate load}}{\text{design ultimate load}} \right)$$

The design ultimate load was calculated by using the ACI 318-63 code formula,

$$P_u = 0.85 f'_c ba + A'_s f_y - A_s f_s$$

where

$P_u = $ design ultimate load

$f'_c = $ maximum 28 day compressive strength of concrete

$b = $ column width

$d = $ column effective depth

$a = $ depth of equivalent rectangular stress blocks

$A'_s = $ area of compressive reinforcement

$A_s = $ area of tensile reinforcement

$f_y = $ yield stress of reinforcement

$f_s = $ stress in tensile reinforcement

The sample statistics of R are

$$\text{mean} = 0.928$$
$$\text{coefficient of variation} = 0.068$$
$$\text{coefficient of skewness} = -0.030$$

Assume that R has a three-parameter gamma PDF. Establish the PDF of R.

If R is assumed to have a three-parameter gamma, then one could calculate the coefficient of skewness for $p(R)$ defined by Eq. (2-49) and then equate it to the sample coefficient of skewness. This solution is left to the reader and involves solving three equations and three unknowns. An alternative and approximate solution follows if the value of c [see Eq. (2-49)], which is the tail limit parameter, is set equal to the mean minus a multiple of the sample standard deviation. With c then known the mean and variance are used to find a and b. For example, consider two cases for calculating c. In the first, c is equal to the mean minus two standard deviations, i.e.,

$$c = \bar{R} - 2\sigma_R = 0.802 \quad \therefore \quad a = 0.0315, b = 4$$

In the second, c is equal to the mean minus four standard deviations; i.e.,

$$c = \bar{R} - 4\sigma_R = 0.676 \quad \therefore \quad a = 0.0158, b = 16$$

This approximate solution can readily be accomplished without a calculator or computer. The reader should plot these two PDF's and compare their shapes.

Example 2-13

A series of tests are conducted on concrete masonry walls. The random variable, denoted R, is

$$R = \left(\frac{\text{test ultimate load}}{\text{allowable compressive load}}\right)$$

where the allowable compressive load is determined from the National Concrete Masonry Associate Code. Six inch width walls were tested. The sample mean was 7.33, and the sample coefficient of variation was 0.21. Assume that R has a three-parameter log-normal PDF, and plot the probability density function. Now, assume that R has a three-parameter gamma PDF, and plot the probability density function. Compare the above plots with the corresponding plots if R had a normal PDF.

The mean and standard deviation of the test data is used to define the normal PDF, which is shown in Fig. Ex. 2-13.

If three-parameter log-normal and gamma are to be used without a knowledge of the sample coefficient of skewness, then as in Example 2-12 an assumption must be made on the location of the tail limit parameter. The following cases are considered and shown in Fig. Ex. 2-13.

Gamma:

$$c = \bar{R} - 2\sigma_R = 4.25 \quad \therefore \quad a = 0.77, b = 4$$
$$c = \bar{R} - 4\sigma_R = 1.17 \quad \therefore \quad a = 0.38, b = 16$$

Log-normal:

$$c = \bar{R} - 2\sigma_R = 4.25 \quad \therefore \quad a = 0.47, b = 1.01$$
$$c = \bar{R} - 4\sigma_R = 1.17 \quad \therefore \quad a = 0.25, b = 1.79$$

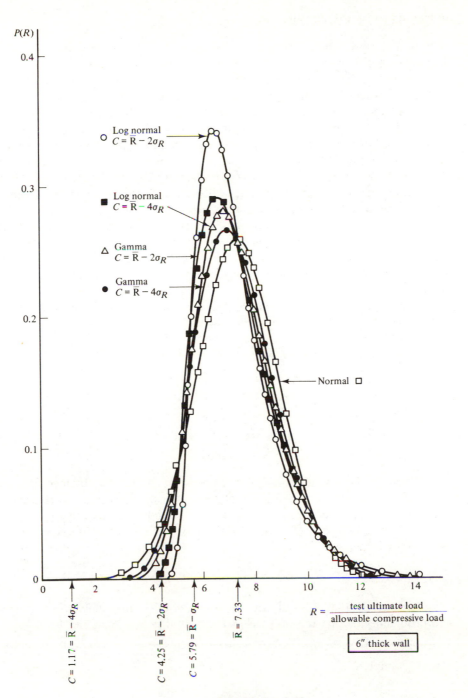

P(R)

Log normal
$C = \overline{R} - 2\sigma_R$

Log normal
$C = \overline{R} - 4\sigma_R$

Gamma
$C = \overline{R} - 2\sigma_R$

Gamma
$C = \overline{R} - 4\sigma_R$

Normal ☐

$C = 1.17 = \overline{R} - 4\sigma_R$

$C = 4.25 = \overline{R} - 2\sigma_R$

$C = 5.79 = \overline{R} - \sigma_R$

$\overline{R} = 7.33$

$R = \dfrac{\text{test ultimate load}}{\text{allowable compressive load}}$

6″ thick wall

FIGURE Ex. 2-13

2-8 CHEBYSHEV INEQUALITY

Selection of a probability density or distribution function and specification of numerical values for its parameters is all that is necessary to calculate the probability that the random variable assumes a numerical value in a prescribed numerical range. Often, in structural engineering, one has a value for the mean and the standard deviation of a random variable and is faced with the problem of specifying, using only a limited amount of data (often intuitive), the probability density function of the random variable. This selection is often an uncomfortable task because the probability associated with any event is a function of the selected probability density function. It is not necessary to select a probability density function to calculate the probability that the random variable will lie in a prescribed numerical range. The probabilistic quantity that is used is the Chebyshev inequality.

The Chebyshev inequality provides an upper-bound estimate of the probability that a random variable will lie outside the prescribed range. However, because one does not have to assume the form of a probability density function, the answers obtained by using the Chebyshev inequality are often too conservative for engineering applications.

The Chebyshev inequality states that the probability that the random variable will be within h standard deviations of the mean is

$$\Pr\left[(\bar{X} - h\sigma_x) \leq X \leq (\bar{X} + h\sigma_x)\right] \geq 1 - \left(\frac{1}{h^2}\right) \tag{2-75}$$

where

$$X = \text{random variable}$$
$$\bar{X} = \text{mean of } X$$
$$\sigma_x = \text{standard deviation of } X$$
$$h = \text{real number}$$

There can be a significant difference between the results obtained by using the Chebyshev inequality and a given probability density function. However, since, in general, one usually does not know the "true" probability density function of a random variable, the assumption of a form for the PDF introduces an uncertainty into any probability estimate.

Example 2-14

Consider Example 2-3. If you use the Chebyshev inequality, then what would the probability be that the steel yield stress would be less than 36.0 ksi?

From Example 2-3

$$\bar{X} = \text{mean yield stress} = 47.9 \text{ psi}$$
$$\sigma_x = \text{standard deviation of yield stress} = 3.3 \text{ ksi}$$

From Chebyshev's inequality the probability that the yield stress will be in the range

$$\bar{X} - h\sigma_x \leq X \leq \bar{X} + h\sigma_x$$

becomes

$$\Pr\,[47.9 - 3.3h \leq X \leq 47.9 + 3.3h] \geq 1 - \left(\frac{1}{h^2}\right)$$

The yield stress value of 36.0 ksi corresponds to an h value of

$$36.0 = 47.9 - 3.3h \quad \therefore \quad h = 3.6$$

Therefore, the probability that the yield stress is within 3.6 standard deviations of the mean is

$$\Pr\,[36.0 \leq X \leq 59.8] \geq 1 - \left(\frac{1}{3.6}\right)^2 = 0.9229$$

The probability that it is outside this range is

$$\Pr\,[X < 36.0 \text{ or } X > 59.8] \leq 1 - 0.9229 = 0.0771$$

Finally, if we assume symmetry,

$$\Pr\,[X < 36.0] \leq \frac{0.0771}{2} = 0.0386$$

2-9 NORMAL AND LOG-NORMAL PROBABILITY DENSITY FUNCTIONS

In Section 2-6 we defined the mathematical form of the normal and log-normal probability density functions. In this section additional consideration is given to these two PDF's because of their special importance in structural engineering applications.

The *normal PDF* is the most commonly used, and even prior to understanding its mathematical definition one has probably heard of its existence (e.g., the "bell-shaped curve"). Equation (2-34) indicates that the PDF is exponential in mathematical form and is symmetric about its mean. The exponential form of the PDF enables it to have many special features which are discussed later in this section.

Equation (2-34) can be written in the form

$$p(x) = \frac{1}{\sigma_x\sqrt{2\pi}} \exp\left\{-\frac{1}{2}\left(\frac{x - \bar{X}}{\sigma_x}\right)^2\right\} \qquad (2\text{-}76)$$

It is convenient to transform Eq. (2-76) into an equivalent but more useful form. Define the new variable

$$U = \left(\frac{X - \bar{X}}{\sigma_x}\right) \qquad (2\text{-}77)$$

and it follows that

$$p(u) = \frac{1}{\sqrt{2\pi}} \exp\left\{-\frac{u^2}{2}\right\} \tag{2-78}$$

Using Eqs. (2-18) and (2-21) it follows that

$$\bar{U} = \text{mean of } U = 0$$

and

$$\sigma_U = \text{variance of } U = 1$$

Thus, the transformation defined by Eq. (2-77) results in a new random variable U which has a normal PDF with zero mean and unit variance. Equation (2-78) is called the *standarized normal PDF*, and Table 2-3 shows a tabulation of the PDF. Figure 2-8 shows the corresponding PDF. The value of $p(x)$ is obtained by dividing the value of $p(u)$ by σ_X, as can be seen from Eqs. (2-76) and (2-78).

TABLE 2-3. Values of the Standardized Normal Probability Density Function

u	0.00	0.01	0.02	0.03	0.04	0.05	0.06	0.07	0.08	0.09
0.0	0.3989	0.3989	0.3989	0.3988	0.3986	0.3984	0.3982	0.3980	0.3977	0.3973
0.1	0.3970	0.3965	0.3961	0.3956	0.3951	0.3954	0.3939	0.3932	0.3925	0.3918
0.2	0.3910	0.3902	0.3894	0.3885	0.3876	0.3867	0.3857	0.3847	0.3836	0.3825
0.3	0.3814	0.3802	0.3790	0.3778	0.3765	0.3752	0.3739	0.3725	0.3712	0.3697
0.4	0.3683	0.3668	0.3653	0.3637	0.3621	0.3605	0.3589	0.3572	0.3555	0.3538
0.5	0.3521	0.3503	0.3485	0.3467	0.3448	0.3429	0.3410	0.3391	0.3372	0.3352
0.6	0.3332	0.3312	0.3292	0.3271	0.3251	0.3230	0.3209	0.3187	0.3166	0.3144
0.7	0.3123	0.3101	0.3079	0.3056	0.3034	0.3011	0.2989	0.2966	0.2943	0.2920
0.8	0.2897	0.2874	0.2850	0.2827	0.2803	0.2780	0.2756	0.2732	0.2709	0.2685
0.9	0.2661	0.2637	0.2613	0.2589	0.2565	0.2541	0.2516	0.2492	0.2468	0.2444
1.0	0.2420	0.2396	0.2371	0.2347	0.2323	0.2299	0.2275	0.2251	0.2227	0.2203
1.1	0.2179	0.2155	0.2131	0.2107	0.2083	0.2059	0.2036	0.2012	0.1989	0.1965
1.2	0.1942	0.1919	0.1895	0.1872	0.1849	0.1826	0.1804	0.1781	0.1758	0.1736
1.3	0.1714	0.1691	0.1669	0.1647	0.1626	0.1604	0.1582	0.1561	0.1539	0.1518
1.4	0.1497	0.1476	0.1456	0.1435	0.1415	0.1394	0.1374	0.1354	0.1334	0.1315
1.5	0.1295	0.1276	0.1257	0.1238	0.1219	0.1200	0.1182	0.1163	0.1145	0.1127
1.6	0.1109	0.1092	0.1074	0.1057	0.1040	0.1023	0.1006	0.0983	0.09728	0.09566
1.7	0.09405	0.09246	0.09089	0.08933	0.08780	0.08628	0.08478	0.08329	0.08183	0.08038
1.8	0.07895	0.07754	0.07614	0.07477	0.07341	0.07206	0.07074	0.06943	0.06814	0.06687
1.9	0.06562	0.06438	0.06316	0.06195	0.06077	0.05959	0.05844	0.05730	0.05618	0.05508
2.0	0.05399									
2.5	0.01753									
3.0	0.00443									
3.5	0.000873									
4.0	0.000134									
5.0	0.00000149									

The standardized normal probability distribution function is

$$P(u_a) = \int_{-\infty}^{u_a} p(u) \, du \qquad (2\text{-}79)$$

and it is shown in Figure 2-8 and tabulated in Table 2-4. The value of the distribution function of X is equal to the corresponding value of the distribution function of U.

It can be shown that a random variable obtained by linearly combining two or more random variables each with a normal PDF has a normal PDF. The mean and variance of the sum are related as shown in Section 2-10.

TABLE 2-4. Values of the Standardized Normal Probability Distribution Function

u	0.00	0.01	0.02	0.03	0.04	0.05	0.06	0.07	0.08	0.09
0.0	0.5000	0.5040	0.5080	0.5120	0.5160	0.5199	0.5239	0.5279	0.5319	0.5359
0.1	0.5398	0.5438	0.5478	0.5517	0.5557	0.5596	0.5636	0.5675	0.5714	0.5753
0.2	0.5793	0.5832	0.5871	0.5910	0.5948	0.5987	0.6026	0.6064	0.6103	0.6141
0.3	0.6179	0.6217	0.6255	0.6293	0.6331	0.6368	0.6406	0.6443	0.6480	0.6517
0.4	0.6554	0.6591	0.6628	0.6664	0.6700	0.6736	0.6772	0.6808	0.6844	0.6879
0.5	0.6915	0.6950	0.6985	0.7019	0.7054	0.7088	0.7123	0.7157	0.7190	0.7224
0.6	0.7257	0.7291	0.7324	0.7357	0.7389	0.7422	0.7454	0.7486	0.7517	0.7549
0.7	0.7580	0.7611	0.7642	0.7673	0.7703	0.7734	0.7764	0.7794	0.7823	0.7852
0.8	0.7881	0.7910	0.7939	0.7967	0.7995	0.8023	0.8051	0.8078	0.8106	0.8133
0.9	0.8159	0.8186	0.8212	0.8238	0.8264	0.8289	0.8315	0.8340	0.8365	0.8389
1.0	0.8413	0.8438	0.8461	0.8485	0.8508	0.8531	0.8554	0.8577	0.8599	0.8621
1.1	0.8643	0.8665	0.8686	0.8708	0.8729	0.8749	0.8770	0.8790	0.8810	0.8830
1.2	0.8849	0.8869	0.8888	0.8907	0.8925	0.8944	0.8962	0.8980	0.8997	0.90147
1.3	0.90320	0.90490	0.90658	0.90824	0.90988	0.91149	0.91309	0.91466	0.91621	0.91774
1.4	0.91924	0.92073	0.92220	0.92364	0.92507	0.92647	0.92785	0.92922	0.93056	0.93189
1.5	0.93319	0.93448	0.93574	0.93699	0.93822	0.93943	0.94062	0.94179	0.94295	0.94408
1.6	0.94520	0.94630	0.94738	0.94845	0.94950	0.95053	0.95154	0.95254	0.95352	0.95449
1.7	0.95543	0.95637	0.95728	0.95818	0.95907	0.95994	0.96080	0.96164	0.96246	0.96327
1.8	0.96407	0.96485	0.96562	0.96638	0.96712	0.96784	0.96856	0.96926	0.96995	0.97062
1.9	0.97128	0.97193	0.97257	0.97320	0.97381	0.97441	0.97500	0.97558	0.97615	0.97670
2.0	0.97725									
2.1	0.98214									
2.2	0.98610									
2.3	0.98928									
2.4	0.99180									
2.5	0.99379									
3.0	0.99865									
3.5	0.999767									
4.0	0.9999683									
4.5	0.9999966									
5.0	0.99999971									
5.5	0.999999981									

u	2.32	3.09	3.72	4.27	4.75	5.20	5.61	6.00	6.36	6.71
$1 - P(u)$	10^{-2}	10^{-3}	10^{-4}	10^{-5}	10^{-6}	10^{-7}	10^{-8}	10^{-9}	10^{-10}	10^{-11}

The symmetry of the normal PDF means that

$$E\langle (X - \bar{X})^n \rangle = 0 \tag{2-80}$$

for $n = 3, 5, \ldots$.
It can be shown that

$$E\langle (X - \bar{X})^n \rangle = \left[\frac{n!}{2^{(n/2)}(n/2)!} \right] \sigma_x^n \tag{2-81}$$

for all even values of n.

The *log-normal PDF* of a random variable is defined by Eq. (2-37). If the random variable X has a log-normal PDF and the random variable Y is defined to be

$$Y = \ln X \tag{2-82}$$

then Y has a normal PDF. Therefore, the PDF of Y can be written as

$$p(y) = \frac{1}{\sigma_Y \sqrt{2\pi}} \exp\left[-\frac{1}{2} \left(\frac{y - \bar{Y}}{\sigma_Y} \right)^2 \right] \qquad -\infty \le y \le +\infty \tag{2-83}$$

It follows upon substitution of Eq. (2-82) into Eq. (2-83) that an alternative form to Eq. (2-37) is

$$p(x) = \frac{1}{x\sigma_Y \sqrt{2\pi}} \exp\left[-\frac{1}{2} \left(\frac{\ln x - \bar{Y}}{\sigma_Y} \right)^2 \right] \qquad x \ge 0 \tag{2-84}$$

The mean of the variable Y can be shown to be equal to

$$\bar{Y} = \ln \bar{X} - \left(\frac{1}{2} \right) \sigma_Y^2 \tag{2-85}$$

and its standard deviation is

$$\sigma_Y^2 = \ln (\rho_X^2 + 1) \tag{2-86}$$

where

$$\rho_X = \text{coefficient of variation} = \left(\frac{\sigma_X}{\bar{X}} \right)$$

Equation (2-84) can now be transformed into standardized normal PDF form by defining

$$U^* \equiv \frac{\ln x - \bar{Y}}{\sigma_Y} \tag{2-87}$$

and

$$p(u^*) = \frac{1}{\sqrt{2\pi}} \exp \left\{ \frac{1}{2} (u^*)^2 \right\} \tag{2-88}$$

It then follows from Eqs. (2-84) and (2-88) that

$$p(x) = \frac{p(u^*)}{x\sigma_Y} \tag{2-89}$$

The distribution function of X and U^* are equal, as was the case for the normal PDF.

Example 2-15

Consider Example 2-11. In that example, the random variable was the ratio of test shear strength to design formula shear strength. The PDF of this random variable was assumed to be a normal and was

$$p(r) = \frac{1}{\sqrt{2\pi}\,(0.167)} \exp\left\{-\frac{1}{2}\left(\frac{r - 1.094}{0.167}\right)^2\right\}$$

If R is less than unity, the design formula is not conservative and overestimates the shear strength capacity. The reader can imagine this as a failure. What is the probability that R will be less than unity?

If the design formula shear strength is divided by a factor of safety of 1.2 and this new shear strength is called the *allowable shear strength*, then what is the probability that the test shear strength will be less than the allowable shear strength?

Recall from Example 2-11 that $\bar{R} = 1.094$ and $\sigma_R = 0.167$. Therefore, an $r = 1$ value corresponds to a U of [see Eq. (2-77)]

$$u = \left(\frac{r - \bar{R}}{\sigma_R}\right) = \frac{1.000 - 1.094}{0.167} = -0.56$$

This means that $r = 1$ is 0.56 standard deviations away from the mean. Because the PDF is symmetric about zero, the probability that u will be less than -0.56 is the same as the probability that it will be greater than $+0.56$. From Table 2-4 this is

$$\Pr[U < -0.56] = \Pr[R < 1] = \Pr[U > +0.56] = 1 - \Pr[U < +0.56]$$
$$= 1 - 0.7123 = 0.29, \text{ or } 29\%$$

The reader is encouraged to make a sketch of the PDF and identify the areas that are being subtracted here. Note that Eq. (2-89) must be used to obtain the PDF.

The allowable shear strength is herein defined to be (see Example 2-11)

$$V_{\text{allowable}} = \left(\frac{V_c}{1.2}\right)$$

where 1.2 is the factor of safety. If the test shear strength is equal to the allowable shear strength this means that

$$\text{test shear strength} = \left(\frac{V_c}{1.2}\right) = 0.833 V_c$$

or that

$$R = \left(\frac{\text{test shear strength}}{\text{design fromula shear strength}}\right)$$

$$= \left(\frac{0.833 V_c}{V_c}\right) = 0.833$$

The probability that R is less than 0.833 is the probability that the test shear strength will be less than the allowable shear strength. Proceeding as before, we obtain

$$u = \frac{r - \bar{R}}{\sigma_R} = \frac{0.833 - 1.094}{0.167} = -1.56$$

and from Table 2-4

$$\Pr\left[U < -1.56\right] = \Pr\left[R < 0.833\right] = \Pr\left[U > +1.56\right]$$

$$= 1 - \Pr\left[U < +1.56\right] = 1 - 0.9406$$

$$= 0.059, \text{ or } 5.9\%$$

Note the decrease in this probability when this factor of safety is used for this example. Factors of safety are discussed in more detail later in this book, but the reader can now begin to appreciate where they can be employed and how important they are in design.

Example 2-16

Consider Examples 2-3 and 2-13. Assume that the yield stress has a normal PDF. Calculate the probability that the yield stress will be less than 36.0 ksi. Then assume that the PDF is a log-normal, and calculate the probability that the yield stress will be less than 36.0 ksi.

From Example 2-3,

$$\bar{X} = \text{mean yield stress} = 47.9 \text{ ksi}$$

$$\sigma_X = \text{standard deviation of yield stress} = 3.3 \text{ ksi}$$

Normally distributed yield stress is

$$u = \frac{x - \bar{X}}{\sigma_X} = \left(\frac{36.0 - 47.9}{3.3}\right) = -3.61$$

From Table 2-4,

$$\text{Pr}\,[U < -3.61] = \text{Pr}\,[X < 36\ \text{ksi}] = 1 - \text{Pr}\,[U > 3.6] = 1 - 0.99981$$
$$= 1.89 \times 10^{-4}$$

Log-normally distributed yield stress:

$$u^* = \frac{\ln x - \bar{Y}}{\sigma_y}$$

$$\sigma_Y = \sqrt{\ln\,(\rho_X^2 + 1)}$$

$$\rho_X = 0.069 \longrightarrow \sigma_Y = 0.0689$$

$$\bar{Y} = \ln \bar{X} - \tfrac{1}{2}\sigma_Y^2 = \ln\,(47.9) - \tfrac{1}{2}(0.00475) = 3.867$$

$$u^* = \frac{\ln\,(36.0) - 3.867}{0.0689} = -4.114$$

$$\text{Pr}\,[U^* < -4.114] = \text{Pr}\,[X < 36\ \text{ksi}] = 1 - \text{Pr}\,[U^* > 4.114]$$
$$= 1 - 0.999975 = 2.6 \times 10^{-5}$$

It is clear that as structural engineers our chances of obtaining a specimen with a yield stress less than 36 ksi are exceedingly small (1.89×10^{-4}) and a log-normal PDF can affect this probability by almost one order of magnitude. Notice how conservative our answers were using Chebyshev's inequality (Example 2-14) compared to the above.

2-10 LINEAR COMBINATION OF RANDOM VARIABLES

In structural engineering, matrix equations often relate one set of variables to another by a linear matrix transformation. The general form of such a transformation is

$$\{Y\} = [C]\ \{X\} \tag{2-90}$$
$$(m \times 1)\ \ (m \times n)(n \times 1)$$

where

$$[C] = (m \times n)\ \text{matrix of constants}$$
$$\{X\} = (n \times 1)\ \text{vector of random variables}$$
$$\{Y\} = (m \times 1)\ \text{vector of random variables}$$

The statistical properties of the $\{Y\}$ random variables can be expressed in terms of the statistical properties of the $\{X\}$ random variables.

Expanding Eq. (2-90) gives the quadratic form

$$Y_j = \sum_{k=1}^{n} C_{jk} X_k \qquad j = 1, 2, \ldots, m \tag{2-91}$$

Taking the expected value of each side of this equation it follows that

$$E\langle Y_j \rangle = \bar{Y}_j = E\left\langle \sum_{k=1}^{n} C_{jk} X_k \right\rangle$$

$$= E\langle C_{j1} X_1 + C_{j2} X_2 + \ldots + C_{jn} X_n \rangle$$

$$= C_{j1} E\langle X_1 \rangle + C_{j2} E\langle X_2 \rangle + \ldots + C_{jn} E\langle X_n \rangle$$

and finally

$$\therefore \quad \bar{Y}_j = \sum_{k=1}^{n} C_{jk} \bar{X}_k \qquad j = 1, 2, \ldots, m \tag{2-92}$$

In matrix notation this is

$$E\langle \{Y\} \rangle = \begin{Bmatrix} E\langle Y_1 \rangle \\ E\langle Y_2 \rangle \\ \vdots \\ E\langle Y_m \rangle \end{Bmatrix} = \{\bar{Y}\}_{(m \times 1)} = [C] \begin{Bmatrix} E\langle X_1 \rangle \\ E\langle X_2 \rangle \\ \vdots \\ E\langle X_n \rangle \end{Bmatrix} = \underset{(m \times n)}{[C]} \underset{(n \times 1)}{\{\bar{X}\}} \tag{2-93}$$

The matrix transformation shown in Eq. (2-93) gives the mean values of the m random variables $(Y_1, Y_2, \ldots, Y_m)$ in terms of the mean values of the n random variables $(X_1, X_2, \ldots, X_n)$.

The covariance matrix of the $\{Y\}$ set of random variables can be expressed in terms of the covariance matrix of the $\{X\}$ set of random variables. To show this consider two random variables Y_j and Y_k. The covariance between these two random variables is the element in the jth row and kth column of the covariance matrix. From Eq. (2-90), it can be seen that

$$Y_j = \sum_{l=1}^{n} C_{jl} X_l \tag{2-94}$$

and

$$Y_k = \sum_{h=1}^{n} C_{kh} X_h \tag{2-95}$$

Using Eqs. (2-92), (2-94), and (2-95) and the definition of the covariance it follows that

$$E\langle (Y_j - \bar{Y}_j)(Y_k - \bar{Y}_k) \rangle = E\left\langle \left(\sum_{l=1}^{n} C_{jl} X_l - \sum_{l=1}^{n} C_{jl} \bar{Y}_l \right) \left(\sum_{h=1}^{n} C_{kh} X_h - \sum_{h=1}^{n} C_{kh} \bar{X}_h \right) \right\rangle$$

$$= E\left\langle \left(\sum_{l=1}^{n} C_{jl}(X_l - \bar{X}_l) \sum_{h=1}^{n} C_{kh}(X_h - \bar{X}_h) \right) \right\rangle$$

$$= E\left\langle \left(\sum_{l=1}^{n} \sum_{h=1}^{n} C_{jl} C_{kh}(X_l - \bar{X}_l)(X_h - \bar{X}_h) \right) \right\rangle$$

$$= \sum_{l=1}^{n} \sum_{h=1}^{n} C_{jl} C_{kh} E\langle (X_l - \bar{X}_l)(X_h - \bar{X}_h) \rangle \tag{2-96}$$

and finally

$$\text{Cov}(Y_j, Y_k) = \sum_{l=1}^{n} \sum_{h=1}^{n} C_{jl} C_{kh} \, \text{Cov}(X_l, X_h) \qquad (2\text{-}97)$$

It can be shown that Eq. (2-97) can be written in the equivalent matrix form,

$$[S_Y] = [C] \quad [S_X] \quad [C]^T \qquad (2\text{-}98)$$
$$(m \times m)\,(m \times n)\,(n \times n)\,(n \times m)$$

where $[S_Y]$ and $[S_X]$ denote, respectively, the covariance matrix of the random variables $\{Y\}$ and $\{X\}$.

Example 2-17

Consider the random variable R to be the ultimate load capacity of a structural member and the random variable S to be ultimate load induced on the structural member. Define a new random variable F as follows:

$$F \equiv R - S$$

If the induced load exceeds the load capacity, failure occurs. Therefore, if F is less than zero, failure occurs. Given the mean and standard deviation of R and S as well as their covariance, calculate the mean and standard deviation of F.

First note that

$$F \equiv R - S \quad \text{or} \quad \{Y\} = [C]\{X\}$$

where

$$\{Y\} = \{F\}$$
$$[C] = [1 \quad -1]$$
$$\{X\} = \begin{Bmatrix} R \\ S \end{Bmatrix}$$

Therefore, from Eq. (2-93) the matrix equation is

$$\{\bar{X}\} = \{\bar{F}\} = [C]\begin{Bmatrix} \bar{R} \\ \bar{S} \end{Bmatrix}$$

or, in scalar form,

$$\bar{F} = \bar{R} - \bar{S}$$

and from Eq. (2-98) the matrix equation for the covariance matrix is

$$[S_Y] = [C]\,[S_X]\,[C]^T$$

or, alternatively,

$$[\text{Var}\,(F)] = [1 \quad -1] \begin{bmatrix} \text{Var}\,(R) & \text{Cov}\,(R, S) \\ \text{Cov}\,(R, S) & \text{Var}\,(S) \end{bmatrix} \begin{bmatrix} 1 \\ -1 \end{bmatrix}$$

In scalar form, the variance of F is

$$\text{Var}\,(F) = \text{Var}\,(R) - 2\,\text{Cov}\,(R, S) + \text{Var}\,(S)$$

Since the mean and standard deviation of F is now known, the probability that it will be less than zero can be directly calculated once a PDF for F is selected. In Chapter 4 we discuss this topic in greater depth.

2-11 REFERENCES AND ADDITIONAL READING MATERIAL

References

[2.1] BENJAMIN, J.R., and C.A. CORNELL (1970): *Probability, Statistics, and Decision for Civil Engineers*, McGraw-Hill Book Company, New York.

[2.2] BURY, K.V. (1975): *Statistical Models in Applied Science*, John Wiley & Sons, Inc., New York.

Additional Reading

FELLER, W. (1960): *An Introduction to Probability Theory and Its Applications*, John Wiley & Sons, Inc. New York.

JOHNSON, M.K., and R.M. LEIBERT (1977): *Statistics*, Prentice-Hall, Inc., Englewood Cliffs, NJ.

ANG, A.H., S. and W.H. TANG (1975): *Probability Concepts in Engineering Planning and Design*, John Wiley & Sons, Inc., New York.

PROBLEMS

2.1   From the concrete data presented in Table 2-2 calculate, using only the first 10 sample values, the following:

     (a) Sample mean, sample variance, and sample coefficient of variation of the secant modulus.

     (b) Sample mean, sample variance, and sample coefficient of variation of the maximum compressive strength.

     (c) Sample covariance and sample correlation coefficient between the secant modulus and maximum compressive strength.

2.2   The following table represents a set of test data obtained for two concentrated loads, P_1 and P_2, acting on a structure.

Sample Number	Random Load Data P₁ (kips)	P₂ (kips)
	P_1 *(kips)*	P_2 *(kips)*
1	2.28	4.57
2	1.61	3.19
3	1.52	2.64
4	1.30	3.32
5	2.52	2.77
6	2.91	3.09
7	2.17	4.49
8	2.63	3.97
9	3.89	3.29
10	1.31	4.48
11	2.86	4.51
12	1.66	4.52
13	3.54	5.06
14	2.54	3.58
15	3.83	4.10
16	2.64	5.11
17	3.75	3.93
18	3.33	3.41
19	3.77	5.31
20	2.22	4.61

(a) Plot the frequency histogram for each random load.

(b) Plot the cumulative frequency histogram for each random load.

(c) Plot the three-dimensional frequency histogram for the two random loads.

(d) Plot the three-dimensional cumulative frequency histogram for the two random loads.

(e) Calculate the sample mean, sample variance, and sample coefficient of variation for each random load.

(f) Calculate the sample covariance and sample correlation coefficient for the two random loads.

2.3 The following table represents a set of test data for a random concentrated load and a random uniformly distributed load acting on a structure.

Sample Number	Random Load Data P_1 *(kips)*	P_2 *(kips/ft)*
1	5.11	0.826
2	7.65	0.705
3	7.04	0.782
4	5.95	0.900
5	7.26	0.961
6	5.31	0.713
7	6.44	0.919
8	6.71	0.862
9	6.07	0.759

| | Random Load Data | |
Sample Number	P_1 (kips)	P_2 (kips/ft)
10	5.93	0.832
11	5.59	0.950
12	7.52	0.755
13	6.74	0.945
14	7.04	0.895
15	5.91	0.753
16	5.93	0.892
17	5.86	0.899
18	6.74	0.757
19	5.91	0.753
20	5.47	0.703

(a) Plot the frequency histogram for each random variable.

(b) Plot the cumulative frequency histogram for each random variable.

(c) Plot the three-dimensional frequency histogram for the two random variables.

(d) Plot the three-dimensional cumulative frequency histogram for the two random variables.

(e) Calculate the sample mean, sample variance, and sample coefficient of variation for each random variable.

(f) Calculate the sample covariance and sample correlation for the two random variables.

2.4 Using the test data in Table 2-2 or the frequency histograms shown in Figs. 2-3 and 2-4 calculate the following:

(a) Probability that the secant modulus of elasticity will be greater than or equal to 3.40×10^6 psi.

(b) Probability that the secant modulus of elasticity will be less than 3.40×10^6 psi.

(c) Probability that the secant modulus of elasticity will be less than 3.70×10^6 psi but equal to or greater than 3.40×10^6 psi.

(d) Probability that the secant modulus of elasticity will be less than 3.70×10^6 psi but equal to or greater than 3.40×10^6 and at the same time the maximum compressive strength will be less than 7.0×10^3 psi.

2.5 Consider the function

$$
\begin{aligned}
p(x_1) &= 0 & x_1 &< 3.2 \times 10^6 \text{ psi} \\
&= c_0(x_1 - 3.2 \times 10^6) & 3.2 \times 10^6 \text{ psi} &\leq x_1 < 3.5 \times 10^6 \text{ psi} \\
&= c_0(3.8 \times 10^6 - x_1) & 3.5 \times 10^6 \text{ psi} &\leq x_1 < 3.8 \times 10^6 \text{ psi} \\
&= 0 & x_1 &\geq 3.8 \times 10^6 \text{ psi}
\end{aligned}
$$

(a) Plot the function vs. x_1.

(b) What must the value of c_0 be in order that the function be a probability density function for x_1.

(c) Calculate the probability that the random variable x_1 will have a value less than 3.3×10^6 psi.

(d) Calculate the probability that the random variable x_1 will have a value between 3.3×10^6 psi and 3.6×10^6 psi.

2.6 Consider the probability density function given by the equation

$$p(x_1) = 0 \qquad\qquad\qquad\qquad x_1 \leq 5 \text{ psf}$$
$$= \frac{1}{20} \exp\left\{-\left(\frac{x_1 - 5}{20}\right)\right\} \qquad x_1 > 5 \text{ psf}$$

(a) Plot the probability density function.

(b) Calculate the probability that the random variable x_1 will have a value between 25 and 30 psf.

2.7 Consider the results of Example 2-6 for the special case where $a = 3.2 \times 10^6$ psi, $b = 3.8 \times 10^6$ psi, $c = 4.0 \times 10^3$ psi, and $d = 10.0 \times 10^3$ psi.

(a) Plot the multivariate (also called *bivariate* or *joint*) probability density function for the two random variables.

(b) Plot the marginal probability density function for each random variable.

(c) Calculate the probability that x_1 will be less than 3.7×10^6 psi but equal to or greater than 3.4×10^6 psi and at the same time x_2 will be less than 7.0×10^3 psi.

2.8 The structural engineering formula for the stress f, at a distance y from the neutral axis of a beam with a moment of inertia I when subjected to a moment M, is

$$f = \frac{My}{I}$$

Consider M to be a random variable and let y and I be deterministic. Calculate the mean, variance, and coefficient of variation of the member stress in terms of mean and variance of the moment M.

2.9 The formula for the fixed end moment of a tapered fixed-fixed beam of length $2l$ with a pin at its center is

$$M = \frac{\omega l^2}{2}$$

Calculate the mean, variance, and coefficient of variation for this moment in terms of the mean and standard deviation of the uniform distributed load ω.

2.10 The maximum moment for a cantilever beam acted upon by a random concentrated load at midspan and a random uniformly distributed load is

$$M = \frac{Pl}{2} + \frac{\omega l^2}{2}$$

where

$$M = \text{maximum moment}$$
$$P = \text{random concentrated load at midspan}$$
$$\omega = \text{random uniformly distributed load}$$
$$l = \text{length of beam}$$

(a) Calculate the mean, standard deviation, and coefficient of variation for this moment M in terms of the mean and standard deviation of P and ω. Assume that P and ω are independent random variables.

(b) Repeat part (a), but now assume that P and ω are correlated random variables with a correlation coefficient equal to ρ.

2.11 Using the probability density function given in Problem 2.5 calculate the mean and coefficient of variation of the random variable x_1.

2.12 Use the multivariate probability density function given in Problem 2.7, and calculate the mean and variance of each random variable and the covariance relating the two random variables.

2.13 If the maximum compressive strength of concrete is considered to be a random variable with a mean equal to 7.3×10^3 psi and a standard deviation equal to 1.37×10^3 psi, then plot its probability density function if:

(a) The probability density function is a uniform.
(b) The probability density function is a normal.
(c) The probability density function is a two-parameter log-normal.
(d) The probability density function is a gamma.
(e) The probability density function is a beta with $a = 4.0 \times 10^3$ psi and $b = 10.0 \times 10^3$ psi.

2.14 In Problem 2.8 the mean and variance of the stress were calculated in terms of M, y, and I. Consider the resultant stress to have a mean of 50.0 ksi and standard deviation of 5.0 ksi. If the stress is assumed to have a uniform probability density function, then plot its probability density function and calculate the probability that the stress will be greater than 47.9 ksi.

2.15 The mean and coefficient of variation for the moment M were calculated in Problem 2.9. If the probability density function of the moment is a normal probability density function, sketch the probability density function.

2.16 In Problem 2.10 consider the equations for the fixed end moment of a beam with a uniform distributed load ω and a concentrated load P.

(a) Assume that the two random variables, ω and P, have a bivariate uniform probability density function. Sketch their bivariate probability density function.

(b) Assume that the two random variables are independent and have a bivariate normal probability density function. Sketch their bivariate probability density function.

2.17 The sample mean and standard deviation for the maximum compressive strength were calculated in Example 2-2. Consider this random variable to have a normal probability density function.

(a) Plot the probability density function for the maximum compressive strength.

(b) Calculate the probability that the maximum compressive strength will be less than 5.0×10^3 psi.

(c) Calculate the probability that the maximum compressive strength will be greater than 8.0×10^3 psi.

(d) Calculate the probability that the maximum compressive strength will be in the range 6.0×10^3 psi to 8.0×10^3 psi.

2.18 Repeat Problem 2.17, but now assume that the maximum compressive strength has a log-normal probability density function.

2.19 The support moment for a fixed-fixed beam acted upon by a random concentrated load at midspan and a random uniformly distributed load is

$$M = \frac{Pl}{8} + \frac{\omega l^2}{12}$$

where

$$P = \text{random concentrated load at midspan}$$

$$\omega = \text{random uniformly distributed load}$$

$$l = \text{length of beam}$$

Let l be deterministic. The midspan deflection for the same problem is

$$\Delta = \frac{Pl^3}{192EI} + \frac{\omega l^4}{384EI}$$

where

$$E = \text{modulus of elasticity}$$

$$I = \text{moment of inertia}$$

Let E and I be deterministic. In Section 2-10 special consideration was given for random variables (e.g., M and Δ) that are linear functions of other random variables (e.g., P and ω).

(a) Define the quantities $\{Y\}$, $[C]$, and $\{X\}$ in Eq. (2-90) for this beam example.

(b) If $E = 30 \times 10^6$ psi, $I = 100$ in.^{4}, and $l = 100$ in., then calculate the vector of mean responses, $\{\bar{Y}\}$, and the covariance matrix, $[S_y]$, for this beam problem in terms of the mean and variances of P and ω and the covariance between P and ω.

2.20   The stiffness or displacement method of structural analysis relates the vector
of forces acting on the structure, $\{F\}$, to the vector of displacements which
result, $\{q\}$, using a stiffness matrix, $[k]$. That is,

$$\{F\} = [k]\{q\}$$

Considering $[k]$ to be deterministic and known, it then follows that

$$\{q\} = [k]^{-1}\{F\}$$

where $[k]^{-1}$ is the inverse of $[k]$. Consider the forces acting on the structure
to be random variables. If the means, variances, and covariances for the
forces are known, discuss how one can calculate the means, variances, and
covariances of the random displacements.

3

Structural Analysis Incorporating Uncertainty

3-1 INTRODUCTION

The rise of matrix methods in structural mechanics has followed the rise of the role of the digital computer in usage and application in industry. Problems that were too complex 20 years ago are now solved routinely on the computer by any one of the large number of available structural programs. This greater flexibility in structural modeling has ushered in two areas which can and must be considered:

1. Accuracy of computation considering numerical round-off and ill-conditioning.
2. Quantification of response uncertainties due to uncertainty in the properties and loads of the structural model.

The first area has been of great importance in implementing computer programs and the subject of considerable study. In this area the structural input is assumed to be precise, and the error to lie only in the numerical computations. The objective of such studies is, then, to contain the errors within set bounds.

The second area assumes that the machine errors are contained but opens for consideration the treatment of the structural properties and loads as random variables.

Structural engineering involves the solution of equations derived from the principles of structural mechanics and utilizes the properties of structural engineering materials. These equations relate structural response quantities (e.g., stress or deflection) to the structural parameters (e.g., loading, moment of inertia, or modulus of elasticity). Therefore, if the structural parameters are random variables, then the structural response quantities are also random variables. In this chapter we show how one can quantify response uncertainties and also make probabilistic statements about prescribed levels of structural response.

Two methods are presented for quantifying structural response uncertainties. The first method uses a Taylor's series expansion to formulate a linear relationship between the response random variables and the random structural parameters. The analysis involves the solution of a system of linear matrix equations and is called a *linear statistical analysis*. The second method involves the use of a computer, or a programmable hand calculator, and simulates an experiment. First, a set of random numbers is generated to represent the statistical uncertainty in the structural parameters. These random numbers are then substituted into the response equation to obtain a set of random numbers which reflect the uncertainty in the structural response. This set is then analyzed by using the techniques discussed in Chapter 2, and a qualification of the uncertainty follows. This type of analysis is called a *Monte Carlo analysis*.

The chapter closes with an introduction to *decision tree analysis*. This method of analysis introduces the reader to the topic of decision theory.

3-2 LINEAR STATISTICAL ANALYSIS

The statistics of structural response which are formulated in this section are based upon a linear statistical model. To understand such a mathematical model consider the notion of a Taylor's series expansion.

Consider a single function which is dependent upon m parameters, $r_1, r_2, \ldots, r_m$. If one defines this function as

$$f(r_1, r_2, \ldots, r_m)$$

then its Taylor's series expansion about a particular set of values of the parameters, say $\bar{r}_1, \bar{r}_2, \ldots, \bar{r}_m$, is

$$f(r_1, r_2, \ldots r_m) = f(\bar{r}_1, \bar{r}_2, \ldots \bar{r}_m) + \sum_{j=1}^{m} \frac{\partial f(\bar{r}_1, \bar{r}_2, \ldots, \bar{r}_m)}{\partial r_j} (r_j - \bar{r}_j) + 0^+ \qquad (3\text{-}1)$$

It is understood that in the summation the partial derivative of the function $f(r_1, r_2, \ldots, r_m)$ is taken and then evaluated at the expansion point $(\bar{r}_1, \bar{r}_2, \ldots, \bar{r}_m)$. Also, 0^+ denotes the higher than linear terms of the Taylor's series expansion, and in a linear statistical model these higher order terms are neglected.

Equation (3-1) is a series expansion of one function. The corresponding matrix equation for n functions of m parameters is

$$\{f(r)\} = \{f(\bar{r})\} + \left[\frac{\partial f(\bar{r})}{\partial r} \right] \{r - \bar{r}\}$$
$$(n \times 1) \qquad (n \times 1) \qquad (n \times m) \ (m \times 1) \qquad\qquad (3\text{-}2)$$

or, in expanded form,

$$\begin{Bmatrix} f_1(r_1, r_2, \ldots r_m) \\ f_2(r_1, r_2, \ldots r_m) \\ \vdots \\ f_n(r_1, r_2, \ldots r_m) \end{Bmatrix} = \begin{Bmatrix} f_1(\bar{r}_1, \bar{r}_2, \ldots, \bar{r}_m) \\ f_2(\bar{r}_1, \bar{r}_2, \ldots, \bar{r}_m) \\ \vdots \\ f_n(\bar{r}_1, \bar{r}_2, \ldots, \bar{r}_m) \end{Bmatrix} +$$

$$\begin{bmatrix} \dfrac{\partial f_1(\bar{r}_1, \bar{r}_2, \ldots, \bar{r}_m)}{\partial r_1} & \cdots & \dfrac{\partial f_1(\bar{r}_1, \bar{r}_2, \ldots, \bar{r}_m)}{\partial r_m} \\ \dfrac{\partial f_2(\bar{r}_1, \bar{r}_2, \ldots, \bar{r}_m)}{\partial r_1} & \cdots & \dfrac{\partial f_2(\bar{r}_1, \bar{r}_2, \ldots, \bar{r}_m)}{\partial r_m} \\ \vdots & & \vdots \\ \dfrac{\partial f_n(\bar{r}_1, \bar{r}_2, \ldots, \bar{r}_m)}{\partial r_1} & \cdots & \dfrac{\partial f_n(\bar{r}_1, \bar{r}_2, \ldots, \bar{r}_m)}{\partial r_m} \end{bmatrix} \begin{Bmatrix} (r_1 - \bar{r}_1) \\ (r_2 - \bar{r}_2) \\ \vdots \\ (r_m - \bar{r}_m) \end{Bmatrix}$$

The partial derivative matrix $\left[\dfrac{\partial f(\bar{r})}{\partial r} \right]$ is referred to as a *sensitivity matrix*.

Now, consider the m parameters $r_1, r_2, \ldots, r_m$ to be a set of random variables. Therefore, the n functions of these parameters, $f_1, f_2, \ldots, f_n$, are also a set of random variables. It may be helpful to imagine, for example, that the random variable r_j, is the modulus of elasticity of the jth structural member, where $f_1, f_2, \ldots, f_n$ are structural displacements. Now, let $\bar{r}_j$ be the *mean of the random variable* r_j, and then Eq. (3-2) is a Taylor's series expansion of each random function about the mean values of the random parameters $r_1, r_2, \ldots, r_m$.

The mean value of each random function is obtained by taking the expected value of both sides of Eq. (3-2)—see Section 2-10— and is

$$E\langle\{f(r)\}\rangle = \begin{Bmatrix} E\langle f_1 \rangle \\ E\langle f_2 \rangle \\ \vdots \\ E\langle f_n \rangle \end{Bmatrix} = \{f(\bar{r})\} + \left[\frac{\partial f(\bar{r})}{\partial r} \right] \begin{Bmatrix} E\langle r_1 - \bar{r}_1 \rangle \\ E\langle r_2 - \bar{r}_2 \rangle \\ \vdots \\ E\langle r_n - \bar{r}_n \rangle \end{Bmatrix}$$

$$= \{f(\bar{r})\} + \left[\frac{\partial f(\bar{r})}{\partial r} \right] E\langle\{r - \bar{r}\}\rangle \tag{3-3}$$

Since each term $E\langle r_j - \bar{r}_j \rangle$ is zero, it follows that

$$E\langle\{f(r)\}\rangle = \{f(\bar{r})\} \tag{3-4}$$

Therefore, to obtain the mean of the functions $f_1, f_2, \ldots, f_n$ one evaluates the functions at the mean values of the random parameters, i.e., $\bar{r}_1, \bar{r}_2, \ldots, \bar{r}_m$.

The covariance matrix of the random functions is obtained by using Eqs. (2-21), (3-2), (3-4), and (2-98). Therefore,

$$[S_f] \equiv \begin{bmatrix} \text{Var}(f_1) & & & & \text{symmetrical} \\ \text{Cov}(f_2, f_1) & \text{Var}(f_2) & & & \cdot \\ & \cdot & & & \cdot \\ & \cdot & & & \cdot \\ & \cdot & & & \cdot \\ \text{Cov}(f_n, f_1) & \text{Cov}(f_n, f_2) & \cdots & \text{Var}(f_n) \end{bmatrix}$$

$$\equiv E\langle (\{f(r)\} - \{f(\bar{r})\})(\{f(r)\} - \{f(\bar{r})\})^T \rangle \tag{3-5}$$

and finally

$$[S_f] = \left[\frac{\partial f(\bar{r})}{\partial r} \right] [S_r] \left[\frac{f(\bar{r})}{\partial r} \right]^T \tag{3-6}$$
$$(n \times n) \quad (n \times m)\,(m \times m)\,(m \times n)$$

where the covariance matrix of the random parameters, r_j, is defined as

$$[S_r] \equiv \begin{bmatrix} \text{Var}(r_1) & & & & \text{symmetrical} \\ \text{Cov}(r_2, r_1) & \text{Var}(r_2) & & & \cdot \\ & \cdot & & & \cdot \\ & \cdot & & & \cdot \\ & \cdot & & & \cdot \\ \text{Cov}(r_m, r_1) & \text{Cov}(r_m, r_2) & \cdots & \text{Var}(r_m) \end{bmatrix}$$

It is often desirable to write Eq. (3-6) in an alternative form which utilizes a correlation coefficient matrix of the random parameters, $r_1, r_2, \ldots, r_m$. Therefore, noting that

$$[S_r] = [\ \sigma_r\ ]\ [\rho]\ [\ \sigma_r\ ] \tag{3-7}$$
$$(m \times m) \quad (m \times m)\,(m \times m)\,(m \times m)$$

where

$$[\ \sigma_r\ ] = \begin{bmatrix} \sigma_{r_1} & 0 & \cdot & \cdot & \cdot & \cdot & 0 \\ 0 & \sigma_{r_2} & 0 & \cdot & \cdot & \cdot & 0 \\ & 0 & & \cdot & \cdot & \cdot & \cdot \\ \cdot & \cdot & \cdot & \cdot & \cdot & \cdot & \cdot \\ \cdot & \cdot & \cdot & \cdot & \cdot & \cdot & \cdot \\ \cdot & \cdot & \cdot & \cdot & \cdot & \cdot & 0 \\ 0 & 0 & \cdot & \cdot & \cdot & 0 & \sigma_{r_m} \end{bmatrix} = \text{standard deviation matrix}$$

and

$$[\rho] = \begin{bmatrix} 1 & \rho_{12} & \rho_{13} & \cdots & \rho_{1m} \\ \rho_{21} & 1 & \rho_{23} & \cdots & \rho_{2m} \\ \cdot & \cdot & \cdot & & \cdot \\ \cdot & \cdot & \cdot & & \cdot \\ \cdot & \cdot & \cdot & & \cdot \\ \rho_{m1} & \rho_{m2} & \rho_{m3} & \cdots & 1 \end{bmatrix} = \text{correlation coefficient matrix}$$

it can be seen that Eq. (3-6) can be written in the alternative form

$$[S] = \left[\frac{\partial f(\bar{r})}{\partial r}\right] [\ \sigma_r\ ]\ [\rho]\ [\ \sigma_r\ ]\left[\frac{\partial f(\bar{r})}{\partial r}\right]^T \qquad (3\text{-}8)$$
$$(n \times n) \quad (n \times m)\ (m \times m)\ (m \times m)\ (m \times m)\ (m \times n)$$

Before we go to the next section, it is important to note that:

1. No assumption is necessary concerning the form of the probability density function of the random parameters, r_j.
2. Knowledge of the mean and covariance matrix of the random structural parameters plus structural response equations is all that is required to obtain the mean and covariance matrix of the response functions.
3. In order to make probabilistic statements concerning response a PDF must be assumed for the response function. For example, Eq. (2-74) can be used if the normal PDF is assumed for the response functions.
4. A check of the accuracy of a linear statistical analysis for a problem where higher-order Taylor's series expansion terms are present can be obtained by a Monte Carlo analysis of the type discussed in the next section.

Example 3-1

A wood cantilever beam weighing 50 lb/ft carries a random upward concentrated force with a mean of 4,000 lb at the end. The coefficient of variation of the upward load is 15%. Calculate the mean and variance of the maximum bending moment at a section 6 ft from the free end.

The moment is

$$M = \text{moment 6 ft from free end}$$

$$= \frac{(50)(6)^2}{2} - 6P = 900 - 6P \text{ lb-ft}$$

where

$$P = \text{upward concentrated force (random variable)}$$

A linear Taylor's series expansion about the mean value of the force gives

$$M = M(P = \bar{P}) + \frac{\partial M}{\partial P}\Big|_{P=\bar{P}} (P - \bar{P})$$

where

$$\bar{P} = \text{mean value of the force} = 4{,}000 \text{ lb}$$

Therefore,

$$M = [900 - 6(4{,}000)] + (-6)(P - 4{,}000)$$
$$= -23{,}100 - 6(P - 4{,}000) \text{ lb-ft}$$

Note that there are no higher than linear terms in the expansion of M, and therefore the solution is exact. The mean moment is

$$\bar{M} = -23{,}100 - 6(4{,}000 - 4{,}000) = -23.1 \text{ kip-ft}$$

and the variance is

$$\sigma_M^2 = \left(\frac{\partial M}{\partial P}\bigg|_{P=\bar{P}}\right)^2 \sigma_P^2 = (36)[(0.15)(4{,}000)]^2 = 12.96 \text{ (kip-ft)}^2$$

Example 3-2

The maximum ground acceleration at a distance of R kilometers from an earthquake of Richter magnitude M can be estimated by using the formula

$$a \text{ (cm/sec}^2) = 1{,}230e^{0.8M}(R + 25)^{-2}$$

Let $M = 6$ and consider it to be deterministic. Consider R to be a random variable with a mean of 25 km and a coefficient of variation of 10%. Use a linear statistical analysis to find the mean, standard deviation, and coefficient of variation of the maximum ground acceleration.

The ground acceleration is

$$a = 1{,}230e^{0.8(6)}(R + 25)^{-2} = C(R + 25)^{-2}$$

where

$$C = 1{,}230e^{0.8(6)} = 149{,}500$$

Using a linear statistical model it follows that

$$a = a(R = \bar{R}) + \frac{\partial a}{\partial R}\bigg|_{R=\bar{R}}(R - \bar{R}) + 0^+$$
$$= C(\bar{R} + 25)^{-2} + (-2C)(\bar{R} + 25)^{-3}(R - \bar{R})$$
$$= 59.8 - 2.39(R - 25)$$

The mean ground acceleration is

$$\bar{a} = E\langle a \rangle = 59.8 \text{ cm/sec}^2$$

and the variance is

$$\sigma_a^2 = \left(\frac{\partial a}{\partial R} \Big|_{R=\bar{R}} \right)^2 \sigma_R^2 = (2.39)^2 [0.10(25)]^2 = 35.7 \ (\text{cm/sec}^2)^2$$

It then follows that the standard deviation and the coefficient of variation of ground acceleration are

$$\sigma_a = \text{standard deviation} = 5.98 \text{ cm/sec}$$

$$\rho = \text{coefficient of variation} = \frac{\sigma_a}{\bar{a}} = \frac{5.98}{59.8} = 10\%$$

Note that since there are higher-order Taylor's series expansion terms, the linear statistical analysis answers are not exact for this problem.

Example 3-3

The wind pressure acting on a window is given by

$$p = \tfrac{1}{2}\rho(V + v)^2$$

where

$$p = \text{pressure}$$
$$\rho = \text{density of air}$$
$$V = \text{basic wind speed}$$
$$v = \text{gust wind speed}$$

Consider ρ and V to be deterministic and v to be random with mean $\bar{v}$ and standard deviation σ_v. Use a linear statistical analysis to find the mean and variance of pressure.

The pressure is

$$p = \tfrac{1}{2}\rho V^2 + \rho V v + \tfrac{1}{2}\rho v^2$$

A linear Taylor's series expansion of this pressure about the mean gust wind speed is

$$p = p(v = \bar{v}) + \frac{\partial p}{\partial v}\Big|_{v=\bar{v}} (v - \bar{v}) + 0^+$$

$$= [\tfrac{1}{2}\rho(V + \bar{v})^2] + [\rho(V + \bar{v})](v - \bar{v}) + 0^+$$

The mean pressure is

$$\bar{p} = \tfrac{1}{2}\rho(V + \bar{v})^2$$

and the variance of the pressure is

$$\sigma_p^2 = \left(\frac{\partial p}{\partial v}\Big|_{v=\bar{v}}\right)^2 \sigma_v^2 = \rho^2(V + \bar{v})^2 \sigma_v^2$$

Example 3-4

The load deflection matrix equation of static equilibrium can be written

$$\underset{(n \times n)}{[K]} \quad \underset{(n \times l)}{\{q\}} \quad = \quad \underset{(n \times l)}{\{F\}}$$

where

$$[K] = \text{stiffness matrix}$$
$$\{F\} = \text{force vector}$$
$$\{q\} = \text{displacement vector}$$

Calculate the mean and covariance matrix of the static deflections.

Using the same general notation for these random variables, $r_1, r_2, \ldots, r_m$, and noting that the random functions under consideration are generalized displacements, one may rewrite Eq. (3-2) as

$$\underset{(n \times 1)}{\{q(r)\}} = \underset{(n \times 1)}{\{q(\bar{r})\}} + \underset{(n \times m)}{\left[\frac{\partial q(\bar{r})}{\partial r}\right]} \underset{(m \times 1)}{\{r - \bar{r}\}}$$

Note that this equation is of the same linear transformation form as discussed in Section 2-10, where

$$[c] = \left[\frac{\partial q(\bar{r})}{\partial r}\right]$$

In this equation $\{q(\bar{r})\}$ is the mean value of the static response and is obtained by solving the load-deflection equation with the mean values of the random variables, r_j, substituted into the stiffness matrix. That is, the mean response is the solution to the matrix equation given by Eq. (2-93) and is

$$E\langle\{q(\bar{r})\}\rangle = \{q(\bar{r})\} = [K]^{-1}\{F\}$$

with $\bar{r}_j$ substituted everywhere for r_j before solving.

The first step for calculating the response covariance matrix—see Eq. (2-98)—involves taking the derivative of both sides of the load-deflection equation with respect to the random variables r_j. So doing, one obtains

$$[K]\frac{\partial}{\partial r_j}\{q\} + \left(\frac{\partial}{\partial r_j}[K]\right)\{q\} = \frac{\partial}{\partial r_j}\{F\}$$

and finally the jth column in the $[c]$ matrix of Eq. (2-98) is

$$\left\{\frac{\partial q(\bar{r})}{\partial r_j}\right\} = [K]^{-1}\left(-\left[\frac{\partial K(\bar{r})}{\partial r_j}\right]\{q(\bar{r})\} + \frac{\partial\{F\}}{\partial r_j}\right)$$
$$(n \times 1) \qquad (n \times n) \qquad (n \times n)\ (n \times 1) \qquad (n \times 1)$$

where

$$\left\{\frac{\partial q(\bar{r})}{\partial r_j}\right\} = \left\{\begin{array}{c} \dfrac{\partial q_1(\bar{r}_1, \bar{r}_2, \ldots, \bar{r}_m)}{\partial r_j} \\[2mm] \dfrac{\partial q_2(\bar{r}_1, \bar{r}_2, \ldots, \bar{r}_m)}{\partial r_j} \\[1mm] \vdots \\[1mm] \dfrac{\partial q_n(\bar{r}_1, \bar{r}_2, \ldots, \bar{r}_m)}{\partial r_j} \end{array}\right\}$$

and

$$\left[\frac{\partial K(\bar{r})}{\partial r_j}\right] = \begin{bmatrix} \dfrac{\partial K_{11}(\bar{r}_1, \bar{r}_2, \ldots, \bar{r}_m)}{\partial r_j} & \cdots & \dfrac{\partial K_{1n}(\bar{r}_1, \bar{r}_2, \ldots, \bar{r}_m)}{\partial r_j} \\[3mm] \dfrac{\partial K_{21}(\bar{r}_1, \bar{r}_2, \ldots, \bar{r}_m)}{\partial r_j} & \cdots & \dfrac{\partial K_{2n}(\bar{r}_1, \bar{r}_2, \ldots, \bar{r}_m)}{\partial r_j} \\[3mm] \vdots & & \vdots \\[3mm] \dfrac{K_{n1}(\bar{r}_1, \bar{r}_2, \ldots, \bar{r}_m)}{\partial r_j} & \cdots & \dfrac{K_{nn}(\bar{r}_1, \bar{r}_2, \ldots, \bar{r}_m)}{\partial r_j} \end{bmatrix}$$

Each of the terms in all matrices has been evaluated, after all appropriate partial differentiation, at the mean values of the random variables $r_1, r_2, \ldots, r_m$.

Now one can completely define the covariance matrix for static response. This matrix equation is

$$[S_q] = \left[\frac{\partial q(\bar{r})}{\partial r}\right] [S_r] \left[\frac{\partial q(\bar{r})}{\partial r}\right]^T$$
$$(n \times n) \quad (n \times m)\ (m \times m)\ (m \times n)$$

where $[S_r]$ is the covariance matrix of the random variables $r_1, r_2, \ldots, r_m$.

3-3 MONTE CARLO ANALYSIS: AN OVERVIEW

Monte Carlo analysis is a powerful engineering tool which enables one to perform a statistical analysis of the uncertainty in structural engineering problems, being particularly useful for complex problems where numerous random variables are related through nonlinear equations. It is often helpful to visualize a Monte Carlo analysis as an experiment which is performed by a computer rather than in a structural engineering laboratory.

The fundamental step in a Monte Carlo analysis is the development of a set of random numbers. These numbers can be mechanically or electronically generated, but today in practice most random number generation is accomplished by using digital computers. A review of the literature at most computer centers will yield a standard computer algorithm for random number generation (e.g., see the Scientific Subroutine Package for IBM computers). Uniform random numbers have the property that for a selected range of values (e.g., 0.0 to 1.0) the generated random number is equally likely to occur anywhere in the range. Stated in terms of the material presented in Sections 2-4 and 2-6, the random numbers have a uniform PDF. Table 3-1 lists uniform random numbers for the range 0 to 100,000 [3.1].

These random numbers can be used for any range of uniform PDF if one scales them properly. For example, if the desired range is 0.0 to 1.0, then they must be scaled by 10^{-5}.

As an example consider the simply supported beam shown in Figure 3-1 with a circular cross section and a uniform loading. The deflection at the midspan of the beam is

$$\Delta = \frac{5\omega l^4}{384EI} \tag{3-9}$$

where

$\Delta =$ deflection at midspan of the beam

$l =$ length of beam

$E =$ modulus of elasticity

$\omega =$ distributed load

$I =$ moment of inertia $= \pi d^4/64$

$d =$ diameter of beam cross section

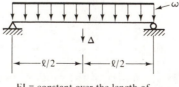

EI = constant over the length of the beam

FIGURE 3-1. Simply supported beam.

TABLE 3-1. Uniform Random Numbers [3.1]†

52478	22835	33307	73842	67277	32880	76457	94489	82597	04836
80249	16089	01964	21414	72117	91712	11487	67479	13649	94539
94132	15190	08425	70298	02202	80519	23516	86294	32871	89573
56605	86696	37707	90117	17511	27701	35764	88217	70505	75300
58815	01919	22225	38562	45731	91743	99315	70350	78240	22015
69379	89366	50240	49343	31867	81661	41037	59120	44282	66605
75228	79546	65528	48794	73980	87645	22604	49290	08068	54935
14327	93484	49875	12103	77984	97966	08644	07089	18809	33738
90625	98430	03639	76657	26389	99093	51145	59343	22488	67026
06070	44497	21962	48270	68632	68338	39325	35105	42348	14412
33415	72559	19902	40024	74215	93857	04988	24389	22094	89237
41999	12790	87990	77646	33177	62684	34119	09212	89973	39638
75908	62356	27342	93069	60284	69329	83998	15037	96165	62149
91323	56853	08468	69550	90860	57946	70370	23114	67185	04633
03428	01736	91578	09165	67708	36704	59481	28243	71395	38607
02333	25192	93932	65485	73266	95972	72606	89242	91968	25721
55696	67106	73369	20689	27707	10432	53118	23692	21450	67362
74838	46105	29798	05504	62588	12700	46093	58754	15780	00361
25833	46204	42441	14284	07858	94467	64358	84445	86230	54172
87260	93170	35494	54207	82683	22976	12257	94522	61364	34228
73595	29104	59346	21213	30923	15747	67104	90389	75901	45606
66224	21746	81973	43832	55932	81707	89193	01511	83257	89931
48078	26348	33935	08981	44947	78208	94370	82235	34382	18908
14168	38881	02968	71715	10814	96338	09439	53864	51951	15691
47813	96995	19524	17227	73490	09448	13156	41802	28217	32658
39404	88593	71327	08978	41241	88350	34760	19507	39102	17168
84131	64236	26803	09167	39695	98995	22498	49489	16808	10807
65097	63684	50298	98391	93703	55438	22718	78013	64409	97879
44552	13101	96263	88862	32977	22191	32112	41046	50771	86355
11997	06462	80215	16900	75972	76712	14861	97496	18986	66671
89716	28633	77208	34231	79158	12531	31612	23543	57480	75667
85258	16576	27023	25722	44809	61284	07636	67054	26665	73238
45790	04380	06893	83032	91230	36690	39612	65695	94966	21734
92386	86028	01737	24812	45158	40744	95550	79951	05457	30445
80321	92435	23677	33356	76405	93136	60668	43458	08562	70311
16964	90116	77618	38200	45273	20442	80655	13676	41471	59063
03060	35414	80332	87759	13961	07849	08970	67354	16026	23225
46517	83209	83758	25428	07686	24628	95824	11554	01428	80580
41481	93999	09645	04406	13666	79199	59323	59115	41436	33185
11580	04688	11925	57414	56554	94938	18151	93058	26924	16181
92862	25355	38189	68819	61797	70112	84563	54657	21490	52086
27419	80915	50829	23146	11641	29047	45806	98176	75455	09782
01450	54579	23503	31250	56057	44450	55982	73182	23666	66578
61200	38309	29934	09351	17290	61419	39377	01770	48134	58599
66047	26430	22415	98215	10413	54380	10492	59665	42368	15138
71899	68860	08150	39941	60556	23386	92449	31012	41277	18925
36567	46306	69777	56251	20007	74448	75234	58915	64903	24311
86135	49654	63467	35906	50560	24921	21109	18652	39797	19964
82155	74998	68901	12964	65056	61967	08628	88194	26741	52840
75099	37473	98759	91653	76447	34010	86452	82362	25185	90842

†This table is reproduced with permission from the RAND Corporation, *A Million Random Digits with 100,000 Normal Deviates*, The Free Press, Glencoe, IL.

Assume that l, E, and ω are deterministic variables and also assume that the diameter is a random variable with a given PDF over the range $d_a \leq d \leq d_b$. How does one study the uncertainty in displacement?

In Section 3-2 we discussed the concept of a linear statistical analysis. Equation (3-9) could be expanded in a Taylor's series and, after truncation of higher-order terms, one could use a linear statistical model to obtain the mean and standard deviation of the tip displacement. Alternatively, the mean and standard deviation of the deflection could be obtained by using the concepts presented in Section 2-5 because the deflection is a function of a random variable (i.e., cross-sectional diameter). However, the above two approaches are not particularly attractive because of the nonlinear relationship between Δ and d. Monte Carlo analysis is desirable for this type of problem; Figure 3-2 shows a schematic of the solution procedure used for such an analysis.

A Monte Carlo analysis involves the generation of one set of n random numbers for each random parameter in the response equation. The response equation is then solved by using each random number in the set. Therefore, the response equation is solved n times; i.e., $i = 1, 2, \ldots, n$. Finally, these values of response are analyzed by

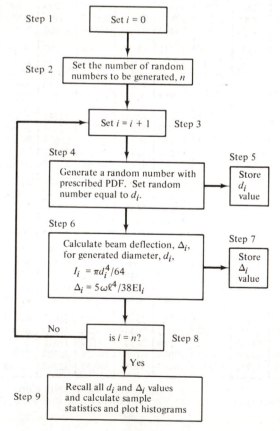

Step 1 — Set $i = 0$

Step 2 — Set the number of random numbers to be generated, n

Step 3 — Set $i = i + 1$

Step 4 — Generate a random number with prescribed PDF. Set random number equal to d_i.

Step 5 — Store d_i value

Step 6 — Calculate beam deflection, Δ_i, for generated diameter, d_i,

$$I_i = \pi d_i^4/64$$

$$\Delta_i = 5\omega l^4/38EI_i$$

Step 7 — Store Δ_i value

Step 8 — is $i = n$?   No / Yes

Step 9 — Recall all d_i and Δ_i values and calculate sample statistics and plot histograms

FIGURE 3-2. *Schematic for the Monte Carlo analysis of a simply supported beam with one random variable.*

using the techniques described in Section 2-3 (i.e., histograms and sample statistics). Because of the nonlinear relationship between beam diameter and midspan displacement, it is not clear what common PDF is most appropriate for describing the displacement. The Monte Carlo analysis directly yields the histogram of displacement and therefore one can (1) make probabilistic statements directly from the histogram or (2) fit a common PDF to the histogram and then use that PDF to make probabilistic statements.

If any of the other parameters on the right-hand side of Eq. (3-9) were random, then the flow chart of Figure 3-2 would be modified to reflect the generation of a set of random numbers for each random variable. In particular, there would be steps analogous to 4 and 5 for each random variable prior to proceeding with the step denoted 6 in Figure 3-2.

Example 3-5

The midspan deflection of a beam subjected to uniform distributed load is given by Eq. (3-9). Assume that four random numbers have been generated for the diameter of the cross section and are

$$d_1 = 5.0 \text{ in.}$$
$$d_2 = 5.8 \text{ in.}$$
$$d_3 = 4.7 \text{ in.}$$
$$d_4 = 5.2 \text{ in.}$$

Use a Monte Carlo analysis, and calculate the mean and standard deviation of the deflection.

The random numbers for the diameter result in the following random values of moment of inertia:

$$d_1 = 5.0 \text{ in.} \longrightarrow I_1 = \frac{\pi(5.0)^4}{64} = 30.68 \text{ in.}^4$$

$$d_2 = 5.8 \text{ in.} \longrightarrow I_2 = 55.55 \text{ in.}^4$$
$$d_3 = 4.7 \text{ in.} \longrightarrow I_3 = 23.95 \text{ in.}^4$$
$$d_4 = 5.2 \text{ in.} \longrightarrow I_4 = 35.89 \text{ in.}^4$$

The corresponding deflections are

$$\Delta_1 = \left(\frac{5\omega l^4}{384E}\right)\left(\frac{1}{I_1}\right) = 4.24 \times 10^{-4}\left(\frac{\omega l^4}{E}\right)$$

$$\Delta_2 = 2.34 \times 10^{-4}\left(\frac{\omega l^4}{E}\right)$$

$$\Delta_3 = 5.44 \times 10^{-4} \left(\frac{\omega l^4}{E}\right)$$

$$\Delta_4 = 3.63 \times 10^{-4} \left(\frac{\omega l^4}{E}\right)$$

Therefore, using Eq. (2-1) the mean deflection is

$$\bar{\Delta} = \left(\frac{\omega l^4}{E}\right)\left(\frac{1}{4}\right)(4.24 + 2.34 + 5.44 + 3.63)(10^{-4}) = 3.91 \times 10^{-4} \left(\frac{\omega l^4}{E}\right)$$

and the variance, using Eq. (2-2) is

$$\text{Var} (\Delta) = 1.25 \times 10^{-8} \left(\frac{\omega l^4}{E}\right)^2$$

and the standard deviation is

$$\sigma_\Delta = 1.12 \times 10^{-4} \left(\frac{\omega l^4}{E}\right)$$

3-4 MONTE CARLO ANALYSIS: INDEPENDENT UNIFORM RANDOM NUMBERS

Table 3-1 lists uniform random numbers that correspond to a PDF—see Eq. (2-31)—with limits $a = 0.00$ and $b = 100,000.00$. These random numbers can be directly transformed into uniform random numbers over the range $a = 0.00$ and $b = 1.00$ if all of the numbers in the table are divided by 10^5 or, alternatively, if a decimal point is placed in front of each number. Therefore, the first two random numbers in column one for an $a = 0.00$ and $b = 1.00$ uniform PDF are 0.52478 and 0.80249, respectively. The later sections of this chapter require uniform random numbers over the range 0 to 1 and therefore this special case is especially important.

Consider the case where one desires uniform random numbers corresponding to an arbitrary uniform PDF. Figure 3-3 shows a plot of the probability distribution function for such a PDF. The solution requires two steps. First, a random number is generated for a uniform PDF with $a = 0.00$ and $b = 1.00$. Second, this random number is transformed to a new random number which corresponds to a uniform PDF over the range a to b. The solution can be illustrated by using Figure 3-3.

Imagine that a random number for a uniform PDF with $a = 0.00$ and $b = 1.00$ has been obtained. Define this random number to be p_i. Locate p_i on the vertical axis of Figure 3-3. Now, construct a horizontal line until it intersects the probability distribution function curve. Then construct a vertical line through that intersection point, and the value of x on the bottom horizontal axis is denoted x_i. That value for x_i is the value of the random number for a uniform PDF over the range a to b.

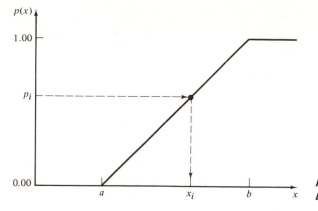

FIGURE 3-3. Uniform proba-bility distribution function.

This solution can be expressed by the equation

$$x_i = a + \left[\frac{b - a}{1.00 - 0.00}\right](p_i - 0.00) \tag{3-10}$$

or, alternatively,

$$x_i = a + (b - a)p_i \tag{3-11}$$

Consider the equation for the probability distribution function for this uniform PDF which results from an integration of Eq. (2-31); i.e.,

$$P(x \leq x_i) = \int_{x=a}^{x=x_i} \left(\frac{1}{b - a}\right) dx = \left(\frac{x_i}{b - a}\right) - \left(\frac{a}{b - a}\right)$$

or, alternatively,

$$x_i = a + (b - a)P(x \leq x_i) \tag{3-12}$$

Therefore, the random number can be obtained from Eq. (3-11) or, alternatively, from Eq. (3-12) with

$$P(x \leq x_i) = p_i$$

This latter method indicates that it is first necessary to integrate the PDF and obtain the corresponding probability distribution function. Then successively solve this equation once for each uniform random within $a = 0.00$ and $b = 1.00$. The value of the desired random number is obtained after this 0 to 1 random number is set equal to the value of the distribution function.

Example 3-6

Consider the list of random numbers shown in Table 3-1. Only the first 20 numbers in column one are to be used in this example.

Generate 20 random numbers with a uniform PDF and a mean of 1.113 and standard deviation of 0.083. This mean and standard deviation corresponds to a random variable that is equal to the ratio of a test plastic bending moment to a design formula plastic moment.

From Eq. (2-32),

$$\bar{X} = 1.113 = \frac{a+b}{2}$$

From Eq. (2-33),

$$\sigma_X^2 = (0.083)^2 = \frac{(b-a)^2}{12}$$

Therefore,

$$a + b = 2.226$$
$$b - a = 0.288$$

and

$$a = 0.969$$
$$b = 1.257$$

Using Eq. (3-10) or (3-11) the first two random numbers are

$$^{(1)}x = 0.969 + \left(\frac{1.257 - 0.969}{1.000 - 0.000}\right)(0.52478 - 0.00000)$$

$$= 1.120$$

$$^{(2)}x = 0.969 + \left(\frac{1.257 - 0.969}{1.000 - 0.000}\right)(0.80249 - 0.00000)$$

$$= 1.200$$

The following table lists the 20 generated random numbers.

Sample Number	Random Number From Table 3-1	Uniform Random Number In Range 0.969 to 1.257
1	52478	1.120
2	80249	1.200
3	94132	1.240
4	56605	1.132
5	58815	1.138
6	69379	1.169
7	75228	1.186
8	14327	1.010
9	90625	1.230
10	06070	0.986
11	33415	1.065
12	41999	1.090

Sample Number	Random Number From Table 3-1	Uniform Random Number In Range 0.969 to 1.257
13	75908	1.188
14	91323	1.232
15	03428	0.979
16	02333	0.976
17	55696	1.129
18	74838	1.185
19	25833	1.043
20	87260	1.220

Example 3-7

Consider the cantilever beam discussed in Example 3-1. Assume that the force has a uniform probability density function. Using the first 20 random numbers in column one of Table 3-1, calculate the 20 corresponding values of the moment 6 ft from the free end of the beam. Calculate the sample mean and variance of the moment.

The mean of the force is

$$\bar{P} = \left(\frac{a + b}{2}\right) = 4.0 \text{ kips}$$

and the variance is

$$\sigma_p^2 = \frac{(b - a)^2}{12} = [(4.0)(0.15)]^2$$

Therefore, the limits on the uniform PDF are

$$a = 2.96 \text{ kips}$$
$$b = 5.04 \text{ kips}$$

The moment corresponding to the ith generated value of the forces is

$$^{(i)}M = 0.90 - 6^{(i)}P$$

Using Eq. (3-11) this becomes, for the first two random numbers,

$$^{(1)}M = 0.90 - 6\left[2.96 + \left(\frac{5.04 - 2.96}{1.00 - 0.00}\right)(0.52478 - 0.00000)\right]$$

$$= -23.41 \text{ kip-ft}$$

$$^{(2)}M = 0.90 - 6\left[2.96 + \left(\frac{5.04 - 2.96}{1.00 - 0.00}\right)(0.80249 - 0.00000)\right]$$

$$= -26.87 \text{ kip-ft}$$

The following table gives the 20 values of random moment.

Sample Number	Random Moment (kip-ft)
1	−23.41
2	−26.87
3	−28.60
4	−23.92
5	−24.20
6	−25.52
7	−26.25
8	−18.65
9	−28.17
10	−17.62
11	−21.03
12	−22.10
13	−26.33
14	−28.25
15	−17.29
16	−17.16
17	−23.81
18	−26.20
19	−20.09
20	−27.75

The sample mean for these 20 values of moment is

$$\bar{M} = -23.7 \text{ kip-ft}$$

and the sample variance is

$$\sigma_M^2 = 14.19 \text{ (kip-ft)}^2$$

Note from Example 3-1 that for this example, where a linear statistical model is exact, the 20-sample Monte Carlo analysis provides approximate but relatively accurate answers.

Example 3-8

Recall from Example 3-2 that the maximum ground acceleration at a distance of R kilometers from an earthquake of Richter magnitude M can be estimated by using the formula

$$a \text{ (cm/sec}^2) = 1{,}230e^{0.8M}(R + 25)^2$$

Let M be a uniform random variable with a mean of 6 and a coefficient of variation of 10%. Consider R to be a uniform random variable with a mean of 25 km and a coefficient of variation of 10%. M and R are assumed to be independent. Use the first 20 numbers in columns one and two of Table 3-1, and perform a Monte Carlo analysis to find the sample mean, sample standard deviation, and sample coefficient of variation of the maximum ground acceleration.

Consider first the Richter magnitude M, where

$$\bar{M} = \text{mean} = 6 = \left(\frac{a_M + b_M}{2}\right)$$

and

$$\sigma_M^2 = \text{variance} = [(0.10)(6)]^2 = \frac{(b_M - a_M)^2}{12}$$

Therefore, the limits on the Richter magnitudes uniform PDF are

$$a_M = 4.96$$
$$b_M = 7.04$$

It similarly follows for R that the limits on its uniform PDF are

$$a_R = 20.67 \text{ km}$$
$$b_R = 29.33 \text{ km}$$

Using Eq. (3-11) the first random numbers for M and R are

$$^{(1)}M = 4.96 + \left(\frac{7.04 - 4.96}{1.00 - 0.00}\right)(0.52478 - 0.00000) = 6.05$$

and

$$^{(1)}R = 20.67 + \left(\frac{29.33 - 20.67}{1.00 - 0.00}\right)(0.22835 - 0.00000)$$
$$= 22.65 \text{ km}$$

Therefore, the first random sample value for the maximum ground acceleration is

$$^{(1)}a = 1{,}230e^{0.80(^{(1)}M)}[^{(1)}R + 25]^2$$
$$= 1{,}230e^{0.80(6.05)}(22.65 + 25)^2$$
$$= 68.60 \text{ cm/sec}^2$$

The following table lists the first 20 sample values for the maximum ground acceleration. The corresponding sample statistics are

$$\text{sample mean} = 71.69 \text{ cm/sec}^2$$
$$\text{sample standard deviation} = 5.69 \text{ cm/sec}^2$$
$$\text{sample coefficient of variation} = 7.9\%$$

Note that the Monte Carlo analysis, not the linear statistical model, provides the exact solution for this problem.

Sample Number	Richter Magnitude Random Value	Earthquake Distance Random Value (km)	Maximum Ground Acceleration (cm/sec²)
1	6.05	22.65	68.60
2	6.63	22.06	111.58
3	6.92	21.99	141.02
4	6.14	28.18	58.99
5	6.18	20.84	82.37
6	6.40	28.41	72.31
7	6.52	27.56	82.30
8	5.26	28.77	28.57
9	6.84	29.19	99.99
10	5.09	24.52	29.35
11	5.66	26.95	42.02
12	5.83	21.78	59.79
13	6.54	26.07	88.16
14	6.86	25.59	116.07
15	5.03	20.82	32.82
16	5.01	22.85	29.55
17	6.11	26.48	61.99
18	6.52	24.66	91.58
19	5.50	24.67	40.53
20	6.77	28.74	96.16

Example 3-9

Consider Example 3-8, where the mean and standard deviation of the maximum ground acceleration have been calculated and are

$\bar{a}$ = mean maximum ground acceleration = 71.69 cm/sec²

σ_a = standard deviation of maximum ground acceleration = 5.69 cm/sec²

A structure exists at the site where this maximum ground motion has been estimated. It has been determined by a structural engineer that:

1. No damage will occur to the structure if the maximum ground acceleration is less than a_1.
2. Moderate damage will occur to the structure if the maximum ground acceleration is between a_1 and a_2.
3. Major structural damage will occur to the structure if the maximum ground acceleration is between a_2 and a_3.

4. Collapse of the structure will occur if the maximum ground acceleration is greater than a_3.

Let $a_1 = 75 \text{ cm/sec}^2$, $a_2 = 100 \text{ cm/sec}^2$, and $a_3 = 125 \text{ cm/sec}^2$. Answer the following questions:

(a) Using the 20 Monte Carlo samples generated in Example 3-8, what is the probability that the structure will have no damage, moderate damage, major damage, or will collapse?

(b) If the maximum ground acceleration is assumed to have a normal PDF, what is the probability that the structure will have no damage? Moderate damage?

Part (a) solution:

A Monte Carlo analysis enables one to review directly the sample results and to identify the structural damage state. For example, sample number (1) maximum ground acceleration is 68.60 cm/sec²; therefore, it is less than a_1 and no damage occurs. Sample number (2) maximum ground acceleration is 111.58 cm/sec²; therefore, it is between a_2 and a_3 and major structural damage occurs. By reviewing the maximum ground acceleration on a sample-by-sample basis, it follows that 11 samples indicate no damage, 6 moderate damage, 2 major damage, and 1 collapse. Therefore, based on the 20-sample Monte Carlo analysis, the following probabilities of damage are obtained:

$$p_1 = \text{probability of no damage} = (11/20) = 55\%$$
$$p_2 = \text{probability of moderate damage} = (6/20) = 30\%$$
$$p_3 = \text{probability of major damage} = (2/20) = 10\%$$
$$p_4 = \text{probability of collapse} = (1/20) = 5\%$$

Note that, for illustrative purposes, the number of samples is only 20. An actual study would use several hundred examples, and therefore a more realistic distribution of samples would result and the calculated probabilities would be more accurate. Also, the structural engineer may wish to be more descriptive in defining damage so that, instead of using only moderate and major damage, one may define more ranges than the four used for this example.

Part (b) solution:

Since the PDF of the maximum ground acceleration has been assumed to be a normal, its PDF is completely defined by using the sample mean and variance. That is,

$$p(a) = \frac{1}{5.69\sqrt{2\pi}} \exp\left\{ -\frac{1}{2}\left(\frac{a - 71.69}{5.69}\right)^2 \right\}$$

The probabilities associated with the damages now follow directly:

$$p_1 = \Pr[a < a_1] = \int_{-\infty}^{a_1} p(a)\, da \quad \text{NO DAMAGE}$$

$$p_2 = \Pr[a_1 \le a < a_2] = \int_{a_1}^{a_2} p(a)\, da \quad \text{MODERATE DAMAGE}$$

Using the material in Section 2-9 it follows that

$$p_1 = 0.719$$
$$p_2 = 0.281$$

Note that if more than 20 samples were used the accuracy of the sample mean and sample variance would improve. Also, since a normal PDF has been assumed for the maximum ground acceleration, a linear statistical analysis could be used to calculate the mean and standard deviation and the rest of the solution would follow the same as discussed in part (b) above.

3-5 MONTE CARLO ANALYSIS: INDEPENDENT RANDOM VARIABLES

The previous section describes the method for generating random numbers for a uniform PDF. Monte Carlo studies would be very limited if they could be used only for uniformly distributed random variables.* Therefore, this section describes how random numbers can be generated for any prescribed PDF. The solution for this problem requires two steps, and the steps are identical in concept to the steps described in the previous section. First, a random number is generated for a uniform PDF with $a = 0.00$ and $b = 1.00$. Second, that random number is transformed to a new random number which corresponds to the prescribed PDF.

Figure 3-3 shows a plot of the probability distribution function for a uniform PDF. One can similarly plot the distribution function corresponding to the prescribed PDF. The second step of the solution directly follows the procedure described in the previous section. That is, the uniform $a = 0.00$ and $b = 1.00$ random number is set equal to p_i, and a horizontal line is drawn until the point of intersection of that line and the distribution function curve is obtained. Then a vertical line is constructed and the value of the random number for the prescribed PDF corresponds to the x_i value on the bottom horizontal axis.

In the previous section it was possible to close form integrate the uniform PDF and obtain a direct relationship between x_i and p_i. This is often possible for a pre-

*Note that many hand calculators have the capability of generating random numbers for several common PDF's. They also often have the capability of calculating sample statistics.

scribed PDF, and therefore one need not always construct a plot of the probability distribution function. Thus, it is often possible to solve for x_i by using

$$P(x = x_i) = \int_{-\infty}^{x_i} p(x)\, dx = p_i \qquad (3\text{-}13)$$

and the corresponding relationship between p_i and x_i.

The digital computer, or alternatively, a hand calculator can be utilized to generate random numbers for a prescribed PDF when one either cannot integrate the PDF or desires not to do so. This is accomplished by representing the probability distribution function corresponding to the prescribed PDF by a sequence of linear lines, as is demonstrated in Figure 3-4 where, for illustration, the probability distribution function has been divided into 10 line segments. The segments in this example divide the vertical axis into 10 equal intervals; however, this is not necessary and these intervals can, in general, be unequal. For the present discussion consider this representation of the probability distribution curve by these m intervals, where $m = 10$.

The vertical (i.e., probability) axis is divided into m intervals of equal probability. These intervals are the length Δp, where

$$\Delta p = \frac{1.0}{m} \qquad (3\text{-}14)$$

Each interval on the probability axis corresponds to an interval on the random variable axis (i.e., p_6 to p_7 corresponds to x_6 to x_7). Figure 3-4 shows the vertical axis

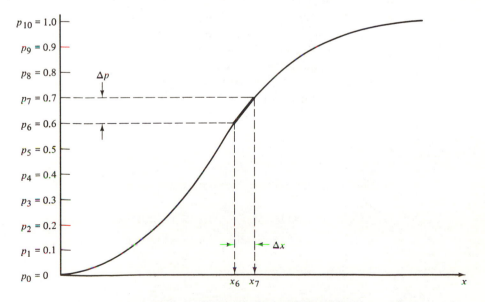

FIGURE 3-4. Cumulative histogram for random number generation.

divided into 10 intervals. It is customary to input to the computer the value of the probability distribution function at $m + 1$ finite points (i.e., $p_j, x_j; j = 0, 1, \ldots, m$). Also, since m is under the control of the engineer performing the random number generation, the value of m is selected to be large enough to ensure that the probability distribution function can be adequately represented by a sequence of input pairs. Therefore, for any value of probability within an interval the value of the random variable can be determined by linear interpolation; i.e., for a value of p_i between p_j and p_{j+1} it follows that

$$x_i = x_j + \left[\frac{x_{j+1} - x_j}{p_{j+1} - p_j}\right](p_i - p_j) \tag{3-15}$$

The generation of random numbers with the plotted probability distribution function is accomplished by generating, using a uniform random number generator, a value for p_i over the interval $a = 0.00$ to $b = 1.00$ and then calculating the corresponding value for the random variable x_i by using Eq. (3-15). Therefore, the steps in this generation are as follows:

Step 1. Plot the probability distribution function of the desired random numbers.

Step 2. Discretize this distribution function into m intervals.

Step 3. Set x_j, x_{j+1} and p_j, p_{j+1} equal to the limits on these intervals. This is done for $j = 0, 1, 2, \ldots, m$.

Step 4. Generate a uniform random number over the range 0 to 1, and identify the interval which the number lies inside on the vertical probability distribution function axis.

Step 5. Use Eq. (3-15) with p_i equal to the random number from step 4, and calculate the random number with the desired PDF. The desired random number is x_i in Eq. (3-15).

Step 6. Repeat steps 4 and 5 until the desired number of random numbers are obtained.

Example 3-10

The PDF of a random variable is given by the equation

$$
\begin{aligned}
p(x) &= 0 & x &< 10 \text{ kips} \\
&= (1/30) & 10 \text{ kips} &\leq x \leq 30 \text{ kips} \\
&= 0 & 30 \text{ kips} &< x < 40 \text{ kips} \\
&= (1/30) & 40 \text{ kips} &\leq x \leq 50 \text{ kips} \\
&= 0 & x &> 50 \text{ kips}
\end{aligned}
$$

Three random numbers have been generated for a uniform PDF with $a = 0.00$ and $b = 1.00$, and they are 0.95, 0.10, and 0.45. Generate the corresponding three random numbers for the above defined PDF.

The probability distribution function corresponding to the prescribed PDF is shown in Fig. Ex. 3-10. Using the generated uniform numbers the random numbers for the prescribed PDF are obtained as shown in Fig. Ex. 3-10.

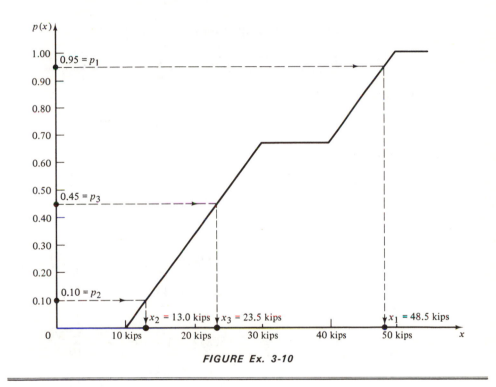

FIGURE Ex. 3-10

Example 3-11

Assume that a random variable has an exponential PDF given by Eq. (2-52) and that its mean and standard deviation are 40 ksi and 10 ksi, respectively. Use the first two random numbers in column one of Table 3-1 and generate two exponential random numbers.

The exponential PDF is

$$p(x) = \frac{1}{a} \exp\left[-\frac{(x - c)}{a}\right] \qquad x > c$$

$$= 0 \qquad \text{otherwise}$$

It follows from Eqs. (2-50) and (2-51) for the mean and variance that

$$\bar{x} = c + a = 40 \text{ ksi}$$
$$\sigma_x^2 = a^2 = 100 \text{ ksi}$$

and therefore

$$a = 10 \text{ ksi}$$
$$c = 30 \text{ ksi}$$

The probability distribution function for the exponential is

$$P(x \leq x_t) = \int_{-\infty}^{x_t} p(x)\, dx = \int_c^{x_t} \frac{1}{a} \exp\left[-\left(\frac{x-c}{a}\right)\right] dx$$

and therefore

$$P(x \leq x_t) = 1 - \exp\left[-\frac{(x_t - 30)}{10}\right]$$

From Table 3-1,

$$p_1 = 0.52478 \quad \text{and} \quad p_2 = 0.80249$$

The two exponential random numbers thus follow and are:

$i = 1$:

$$P(x \leq x_1) = p_1 = 0.52478 = 1 - \exp\left[-\frac{(x_1 - 30)}{10}\right]$$

$$\therefore \quad x_1 = 37.4 \text{ ksi}$$

$i = 2$:

$$p(x \leq x_2) = p_2 = 0.80249 = 1 - \exp\left[-\frac{(x_2 - 30)}{10}\right]$$

$$\therefore \quad x_2 = 46.2 \text{ ksi}$$

Example 3-12

Consider the cantilever beam discussed in Example 3-1. However, now assume that the weight and concentrated load are both independent normally distributed random variables with the following properties:

$$\bar{P} = \text{mean upward concentrated load} = 4{,}000 \text{ lb}$$
$$\sigma_p = \text{standard deviation of upward concentrated load} = 400 \text{ lb}$$
$$\omega = \text{mean beam weight per foot} = 50 \text{ lb/ft}$$
$$\sigma_\omega = \text{standard deviation of beam weight per foot} = 5 \text{ lb/ft}$$

Generate five random numbers for each random variable. Then perform a Monte Carlo analysis and calculate the mean and standard deviation of the moment 6 ft from the free end.

The transformed standard normal PDF is

$$p(u) = \frac{1}{\sqrt{2\pi}} \exp \left\{ -\frac{u^2}{2} \right\}$$

where

$$U = \frac{X - \bar{X}}{\sigma_X}$$

From Table 3-1 the uniform random numbers for P and ω are (0.52478, 0.80249, 0.94132, 0.56605, 0.58815) and (0.69379, 0.75228, 0.14327, 0.90625, 0.06070), respectively.

Table 2-4 is now used for each of these uniform random numbers to obtain the corresponding values of the standard normal u value, which is accomplished by equating each uniform random number to a corresponding standardized probability distribution value. That is,

$$0.52478 = \Pr [u \leq \hat{u}], \quad \text{where} \quad \hat{u} = 0.06220$$

It therefore follows that

$$
\left.
\begin{aligned}
0.52478 &\longrightarrow 0.06220 \\
0.80249 &\longrightarrow 0.85068 \\
0.94132 &\longrightarrow 1.56598 \\
0.56605 &\longrightarrow 0.16628 \\
0.58815 &\longrightarrow 0.22269
\end{aligned}
\right\} \text{ for } P \text{ determination}
$$

and

$$
\left.
\begin{aligned}
0.69379 &\longrightarrow 0.49830 \\
0.75228 &\longrightarrow 0.68181 \\
0.14327 &\longrightarrow -1.06578 \\
0.90625 &\longrightarrow 1.31804 \\
0.06070 &\longrightarrow -1.54893
\end{aligned}
\right\} \text{ for } \omega \text{ determination}
$$

The transformation equation relating U and X is now used to find the random numbers for P and ω. That is,

$$U = \frac{X - \bar{X}}{\sigma_X} \Longrightarrow P = 4{,}000 + 400u$$

$$\omega = 50 + 5u$$

Therefore,

Sample 1:
$$P = 4,000 + 400(0.06220) = 4,024.88$$
$$\omega = 50 + 5(0.49830) = 52.49$$
$$M = 18\omega - 6P = 18(52.49) - 6(4,024.88)$$
$$= -23.20 \text{ kip-ft}$$

Sample 2: $P = 4,340.27,\quad \omega = 53.41,\quad M = -25.08 \text{ kip-ft}$

Sample 3: $P = 4,626.39,\quad \omega = 44.67,\quad M = -26.95 \text{ kip-ft}$

Sample 4: $P = 4,066.51,\quad \omega = 56.59,\quad M = -23.38 \text{ kip-ft}$

Sample 5: $P = 4,089.08,\quad \omega = 42.26,\quad M = -23.77 \text{ kip-ft}$

Finally, the sample mean is $\bar{M} = -24.48$ kip-ft, and the sample standard deviation is $\sigma_M = 1.40$ kip-ft.

3-6 MONTE CARLO ANALYSIS: CORRELATED NORMAL RANDOM VARIABLES

The generation of random numbers for statistically independent random variables is a direct extension of the method used to generate and perform a single random variable Monte Carlo analysis. Such an analysis was actually performed in Examples 3-8 and 3-12. However, it may arise that the structural engineer cannot assume that the random variables are statistically independent. Therefore, in this section we describe how correlated random numbers can be generated for the special case where the joint probability density function of the random variables are normally distributed.

Consider n correlated random variables $(X_1, X_2, \ldots, X_n)$ with a joint normal probability density function. The PDF is defined if the means, variances, and covariance of the random variables are known. Herein, it is assumed that the structural engineer has established values for these and therefore that the mean vector and covariance matrix are known. Thus, one can define the mean vector

$$\{\bar{X}\} = \begin{Bmatrix} \bar{X}_1 \\ \bar{X}_2 \\ \cdot \\ \cdot \\ \cdot \\ \bar{X}_n \end{Bmatrix}$$

and the covariance matrix

$$[S_X] = \begin{bmatrix} \text{Var}(X_1) & \text{Cov}(X_1, X_2) & \cdots & \text{Cov}(X_1, X_n) \\ \text{Cov}(X_2, X_1) & \text{Var}(X_2) & \cdots & \text{Cov}(X_2, X_n) \\ \cdot & \cdot & & \cdot \\ \cdot & \cdot & & \cdot \\ \cdot & \cdot & & \cdot \\ \text{Cov}(X_n, X_1) & \text{Cov}(X_n, X_2) & \cdots & \text{Var}(X_n) \end{bmatrix}$$

Assume that one desires to generate m random numbers for each random variable. The jth generated random number for the ith random variable is denoted as

$$^{(j)}x_i \qquad \begin{array}{l} j = 1, 2, \ldots, m \\ i = 1, 2, \ldots, n \end{array}$$

The random number generation process starts by first generating n sets of statistically *independent* normally distributed random numbers with a yet to be specified mean vector and convariance matrix. Each set has m random numbers. These random numbers are denoted as

$$^{(j)}y_i \qquad \begin{array}{l} j = 1, 2, \ldots, m \\ i = 1, 2, \ldots, n \end{array}$$

A linear transformation is then defined which couples the Y random variables in such a way as to form the correlated X random variables. That is,

$$
\begin{aligned}
^{(j)}x_i &= \sum_{k=1}^{n} C_{ik}^{(j)} y_k \\
&= C_{i1}^{(j)} y_1 + C_{i2}^{(j)} y_2 + \ldots + C_{in}^{(j)} y_n \qquad \begin{array}{l} j = 1, 2, \ldots, m \\ i = 1, 2, \ldots, n \end{array}
\end{aligned} \tag{3-16}
$$

or, alternatively for each j,

$$
\underset{(n \times 1)}{\{^{(j)}x\}} = \underset{(n \times n)}{[C]} \ \underset{(n \times 1)}{\{^{(j)}y\}} \tag{3-17}
$$

In Section 2-10 it is shown that if two sets of random variables are related by a linear transformation, e.g.,

$$\{X\} = [C]\{Y\} \tag{3-18}$$

then

$$\{\bar{X}\} = [C]\{\bar{Y}\} \tag{3-19}$$

and

$$[S_x] = [C][S_y][C]^T \tag{3-20}$$

where

$$\{\bar{X}\}, \{\bar{Y}\} = \text{mean vectors}$$

$$[S_x], [S_y] = \text{covariance matrices}$$

Note that $\{\bar{X}\}$ and $[S_x]$ have the properties of the random variables one wishes to generate and are therefore known. One must find $[C]$, $\{\bar{Y}\}$, and $[S_y]$ such that sets of independent normal random numbers, $^{(j)}y_k$, can be transformed by using Eqs. (3-16) or (3-17) into sets of correlated normal random numbers, $^{(j)}x_i$, with their mean vector and covariance matrix given by $\{\bar{X}\}$ and $[S_x]$, respectively. These matrices are found by using the Choleski decomposition method [3.2].

The Choleski decomposition method is useful in several areas of structural mechanics. The method states that a symmetric matrix of order $n \times n$, denoted $[A]$, can be decomposed into the matrix product

$$[A] = [V][D][V]^T \tag{3-21}$$

$[V]$ is a lower triangular matrix of order $n \times n$ with ones on the principal diagonal, and $[D]$ is a diagonal matrix of order n. The elements of these matrices are

$$D_{11} = A_{11}$$
$$V_{ii} = 1 \qquad i = 1, 2, \ldots, n$$
$$V_{j1} = \frac{A_{1j}}{D_{11}} \qquad j \geq 2$$
$$D_{ii} = A_{ii} - \sum_{l=1}^{i-1} V_{il}^2 D_{ll} \qquad i \geq 2$$
$$V_{ji} = \frac{1}{D_{ii}} \left[A_{ij} - \sum_{l=1}^{i-1} V_{il} V_{jl} D_{ll} \right] \qquad i \geq 2, j \geq i+1$$

It can be seen by inspection of Eqs. (3-20) and (3-21) that

$$[A] = [S_x]$$
$$[C] = [V] \quad \text{(lower triangular matrix)}$$
$$[S_Y] = [D] \quad \text{(diagonal matrix)}$$

and therefore the appropriate linear transformation matrix $[C]$ and covariance matrix $[S_Y]$ are now defined. *The diagonal covariance matrix implies independence of the Y random variables.* The mean vector $\{\bar{Y}\}$ follows directly from

$$\{\bar{X}\} = [C]\{\bar{Y}\} \tag{3-22}$$

by direct Gaussian elimination due to the lower triangular form of the $[C]$ matrix. The $\{X\}$ random variables are normally distributed because they are obtained by a linear transformation of a set of normal random numbers.

In summary, the generation of n sets of normally distributed correlated random numbers, each set with m random numbers, is accomplished by solving Eq. (3-17) m times. The values of the elements in the $[C]$ matrix are obtained by decomposing the covariance matrix of the desired random numbers by using Eq. (3-21) with $[A] = [S_x]$ and $[V] = [C]$. The values of the elements in the $\{^{(j)}y\}$ vector are obtained, for each j, $j = 1, 2, \ldots, m$, by generating n independent normally distributed random variables. The means of these independent random variables are obtained by using Eq. (3-22), and their diagonal covariance matrix is obtained by using Eq. (3-21) with $[S_Y] = [D]$. Therefore, independent normally distributed random numbers are generated, and after an appropriate linear transformation they become correlated normally distributed random numbers. The random generation of correlated random variables that are not normally distributed is beyond the scope of this book.

Example 3-13

In Example 3-12 we performed a Monte Carlo analysis to determine the mean and standard deviations of the load-induced moment. In that example the concentrated load and the weight are assumed to be independent random variables. Now assume that the concentrated load and the weight are correlated normally distributed random variables with a correlation coefficient equal to 0.30. Generate five random numbers for each random variable. Then perform a Monte Carlo analysis, and calculate the mean and standard deviation of the moment 6 ft from the free end.

Recall that

$$\bar{P} = 4,000 \text{ lb}, \quad \sigma_p^2 = 160,000 \text{ lb}^2$$
$$\bar{\omega} = 50 \text{ lb/ft}, \quad \sigma_\omega^2 = 25 \text{ lb}^2/\text{ft}^2$$

and

$$\text{Cov}(p, \omega) = 0.3\sigma_p\sigma_\omega = 600 \text{ lb}^2/\text{ft}$$

It therefore follows that

$$[S_X] = \begin{bmatrix} \sigma_p^2 & \text{Cov}(p, \omega) \\ \text{Cov}(p, \omega) & \sigma_\omega^2 \end{bmatrix} = \begin{bmatrix} 160,000. & 600. \\ 600. & 25. \end{bmatrix}$$

This decomposes [see Eq. (3-21)] to

$$[S_X] = [C][S_Y][C]^T$$

where

$$[C] = \begin{bmatrix} 1 & 0 \\ 0.00375 & 1 \end{bmatrix}$$

and

$$[S_Y] = \begin{bmatrix} 160,000 & 0 \\ 0 & 22.75 \end{bmatrix}$$

Therefore, Eq. (3-22) becomes

$$\{\bar{X}\} = [C]\{\bar{Y}\}$$

where

$$\{\bar{X}\} = \begin{Bmatrix} 4,000 \\ 50 \end{Bmatrix} = \begin{bmatrix} 1 & 0 \\ 0.00375 & 1 \end{bmatrix} \begin{Bmatrix} \bar{Y}_1 \\ \bar{Y}_2 \end{Bmatrix}$$

and then

$$\begin{Bmatrix} \bar{Y}_1 \\ \bar{Y}_2 \end{Bmatrix} = \begin{Bmatrix} 4{,}000 \\ 35 \end{Bmatrix}$$

Using Tables 3-1 and 2-4 and Eq. (3-15) the independent random variables are

Sample 1: $y_1 = 4{,}024.88, \quad y_2 = 35.83$

Sample 2: $y_1 = 4{,}340.27, \quad y_2 = 36.13$

Sample 3: $y_1 = 4{,}626.39, \quad y_2 = 33.23$

Sample 4: $y_1 = 4{,}066.51, \quad y_2 = 37.29$

Sample 5: $y_1 = 4{,}089.08, \quad y_2 = 32.43$

Using Eq. (3-17) it follows that

Sample 1: $P = 4{,}024.88, \quad \omega = 50.92 \quad$ and $\quad M = -23.23$ kip-ft

Sample 2: $P = 4{,}340.27, \quad \omega = 52.41 \quad$ and $\quad M = -25.10$ kip-ft

Sample 3: $P = 4{,}626.39, \quad \omega = 50.58 \quad$ and $\quad M = -26.85$ kip-ft

Sample 4: $P = 4{,}066.51, \quad \omega = 52.54 \quad$ and $\quad M = -23.45$ kip-ft

Sample 5: $P = 4{,}089.08, \quad \omega = 47.76 \quad$ and $\quad M = -23.68$ kip-ft

Finally, the sample mean is $\bar{M} = -24.46$ kip-ft, and the sample standard deviation is $\sigma_M = 1.36$ kip-ft.

3-7 DECISION TREE ANALYSIS

Structural engineers make decisions in the face of uncertainty. These decisions involve their individual firm or company, their clients, and their profession in general when they serve on code or standard preparation committees. Therefore, in this section we present an introduction to the field of decision-making.

Consider for illustration a simple example. The owner of a building must make a decision as to whether to obtain earthquake insurance on the building. The duration of the insurance policy is 3 years. Uncertainty exists in the decision-making process because there is only a chance or probability that a major destructive earthquake will occur during the duration of the policy. Therefore, a decision must be made concerning this uncertainty.

The decision-maker must first identify the possible *courses of action*. In this example, one can either purchase earthquake insurance or not. Therefore, the courses of action are:

a_1 = take out earthquake insurance

a_2 = do not take out earthquake insurance

The next task is to identify all of the possible *states of nature*. Consider herein that only two possible states of nature exist and that they are:

s_1 = no destructive earthquake will occur during the 3 year policy period

s_2 = a destructive earthquake will occur during the 3 year policy period

Each conjunction of an action chosen by the decision-maker and a state of nature, chosen by nature, results in a *consequence*. Therefore, the following four possible consequences exist:

C_{11} = insurance is purchased and no destructive earthquake occurs

C_{12} = insurance is purchased and a destructive earthquake occurs

C_{21} = insurance is not purchased and no destructive earthquake occurs

C_{22} = insurance is not purchased and a destructive earthquake occurs

The consequences are represented by C with a double subscript. The first subscript denotes the action taken, and the second subscript denotes the state of nature.

The decision-maker will react to the possible consequences in different ways. In decision problems it is necessary to rank and quantify one's degree of preference for each possible consequence. This is accomplished by using as a measure of that preference a term known as *utility*. One quantifies the relative utility associated with each consequence by assigning a numerical value to the utility associated with each consequence. In this example a scale is chosen for utility, ranging from 0 to 10. The actual numerical scale is not important as long as one uses the same scale for measuring preference for all consequences. A utility associated with the top of the scale, in this example 10, is assigned to the most preferable consequences—insurance is not taken out and no destructive earthquake occurs. A utility of 0, the value at the bottom of the scale, is assigned to the least preferable consequence—insurance is not taken out and a destructive earthquake occurs. The assignment of utility values to the other consequences will usually be very dependent upon the background and goals of the decision-maker, which are discussed in more depth later in this section. However, in keeping with the illustrative nature of this example, a utility value of 6 is assigned to the consequence that insurance is taken out and a destructive earthquake occurs. In summary, the assigned utilities are as follows:

$$u_{11} = 8$$

$$u_{12} = 6$$

$$u_{21} = 10$$

$$u_{22} = 0$$

where the utilities are represented by u with a double subscript. The first subscript denotes the action taken, and the second subscript denotes the state of nature.

The decision-maker must now assign *probabilities of occurrence* to the different states of nature. These probabilities quantify the possibility that a destructive earthquake will occur during the duration of the earthquake policy. The probabilities of occurrence are denoted as

$$p_1 = \text{probability that state of nature 1 occurs}$$

$$p_2 = \text{probability that state of nature 2 occurs}$$

The sum of these probabilities of occurrence must be unity because the decision-maker has previously identified all possible states of nature. In this illustrative example assume that the states of nature are equally likely to occur, and therefore $p_1 = 0.50$ and $p_2 = 0.50$.

Consider now the utility, or degree of preference, associated with purchasing earthquake insurance. Either an earthquake does not occur (a probability of 0.50) with a utility value of 8, or it does occur (a probability of 0.50) with a utility value of 6. Since one does not know if the earthquake will occur (i.e., which state of nature will actually exist), one calculates the *expected utility* associated with buying earthquake insurance. The expected utility associated with action 1 is denoted E_1 and is equal to

$$E_1 = u_{11}p_1 + u_{12}p_2$$
$$= (8)(0.50) + (6)(0.5) = 7 \qquad (3\text{-}23)$$

Similarly, the expected utility associated with action 2 (i.e., do not buy earthquake insurance) is denoted E_2 and is equal to

$$E_2 = u_{21}p_1 + u_{22}p_2 = 5 \qquad (3\text{-}24)$$

The optimal action in this case, where the decision is based on maximizing the expected utility to the decision-maker, is action 1.

In many decision problems there are more than two possible actions and two possible states of nature. Therefore, Eqs. (3-23) and (3-24) can be represented in the summation form

$$E_i = \sum_{j=1}^{n} u_{ij}p_j \qquad i = 1, 2, \ldots, m \qquad (3\text{-}25)$$

where

$$E_i = \text{expected utility of action } i$$

$$u_{ij} = \text{utility associated with the consequence that action } i \text{ is taken}$$
$$\text{and state of nature } j \text{ occurs}$$

$$p_j = \text{probability of occurrence of state of nature } j$$

$$n = \text{number of possible states of nature}$$

$$m = \text{number of possible actions}$$

It directly follows that Eq. (3-25) can be written as the matrix equation

$$\underset{(m \times 1)}{\{E\}} = \underset{(m \times n)}{[u]} \ \underset{(n \times 1)}{\{p\}} \tag{3-26}$$

When the number of possible actions and possible states of nature become large, *decision trees* are used to clearly and logically carry out the operations defined by Eqs. (3-25) and (3-26). Figure 3-5 shows a decision tree for the illustrative problem just discussed. Note that two types of symbols are used: squares and circles. The square represents a choice made by the decision-maker and is called a *decision node*.

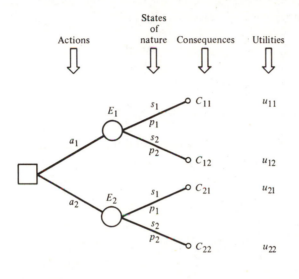

FIGURE 3-5. Decision tree for illustrative example.

The circle represents a choice made by nature and is called a *random node*. Lines branching off from a decision node represent possible actions the decision-maker can take, and lines branching off from a random node represent possible states of nature. The possible consequences and associated utilities are indicated at the end of the random node branches. The probabilities of occurrence associated with each state of nature can readily be identified because they are placed directly beneath the appropriate random node branch. The expected utility associated with each action is identified on the tree and is placed directly above the appropriate random node.

The expected utility associated with an action can be obtained by summing over all branches emanating from a random node the probability for that branch times the utility associated with the consequence identified at the end of the branch. The decision-maker selects the action corresponding to the branch emanating from the decision node which has the largest expected utility noted at its end.

The utility assigned to each consequence in a decision-making problem has a fundamental influence in the action selected by the decision-maker. It is for this reason that different decision-makers given the same decision alternatives, states of nature, and associated probabilities of occurrences will select different optimal actions. Since it is common to indirectly use dollar values to arrive at utility rankings associated with different consequences, a few introductory comments about the establishment of utilities is appropriate.

Imagine that in the previous example the decision-maker can assign dollar values associated with each of the possible consequences. Specifically, imagine that the following dollar/consequence relationships apply:

$$c_{11} \Longrightarrow -\$40,000$$
$$c_{12} \Longrightarrow -\$80,000$$
$$c_{21} \Longrightarrow \quad \$0$$
$$c_{22} \Longrightarrow -\$200,000$$

If utilities are again assigned on a 0 to 10 scale and they are linearly related to dollars, it follows that the corresponding utility values are

$$u_{11} = 8$$
$$u_{21} = 6$$
$$u_{12} = 10$$
$$u_{22} = 0$$

These utilities are the same as previously discussed, and therefore the decision-maker would select action 1. Figure 3-6 shows this linear relationship between dollars and utility. If another decision-maker wishes to select the optimal action for this example problem but, instead of having a linear dollar/consequence relationship a bilinear relationship exists of the form shown in Figure 3-6, the decision-maker would have a different set of utilities. These utilities are

$$u_{11} = 6$$
$$u_{21} = 2$$
$$u_{12} = 10$$
$$u_{22} = 0$$

Therefore, the expected utilities associated with actions 1 and 2 are now equal to

$$E_1 = (6)(0.50) + (2)(0.50) = 4$$

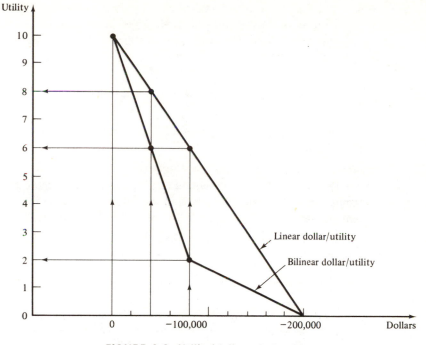

FIGURE 3-6. Utility/dollar relationships.

and

$$E_2 = (10)(0.50) + (0)(0.50) = 5$$

The decision-maker would select action 1 for a linear dollar/utility relationship and action 2 for a bilinear dollar/utility relationship.

The preceding example, while idealistically simple, does illustrate the following points:

1. All possible actions must be identified.
2. The possible states of nature must be identified and structural engineering input is required to define the state of nature in terms meaningful for assignment of probabilities of occurrence and utility.
3. The utility associated with possible action/state of nature consequences must be developed carefully and usually is based on dollar loss information. Structural engineering input is essential at this point because only the structural engineer has the expertise to quantify the relationship between the existence of a state of nature and the dollar impact on the structure.
4. Probabilities of occurrence must be established constant with the defined states of nature. Therefore, one must continually improve one's description of the state of nature. However, one must be aware of one's confi-

dence in assigning realistic probabilities of occurrence for the defined states of nature. It is often the case that the most desirable description of the state of nature goes beyond one's ability in practice to confidently predict its occurrence.

Example 3-14

Consider Example 3-9 and assign the following losses associated with the identified damage states:

Damage state 1: No damage and no loss.
Damage state 2: Moderate damage and a loss of 20% of the building's cost.
Damage state 3: Major damage and loss of 50% of the building's cost.
Damage state 4: Collapse and a loss of 80% of the building's cost.

Calculate the expected loss to the structure, associated with the maximum ground acceleration information given in Example 3-8.

From Examples 3-8 and 3-9(a) it follows that:

Damage State	Loss	Probability	Expected Loss
1	0.0	0.55	0.00
2	0.2	0.30	0.06
3	0.5	0.10	0.05
4	0.8	0.05	0.04

Therefore, the total expected loss is

$$0.00 + 0.06 + 0.05 + 0.04 = 0.15$$

3-8 REFERENCES AND ADDITIONAL READING

References

[3.1] RAND Corporation (1955): *A Million Random Digits with 100,000 Normal Deviates*, The Free Press, Glencoe, IL.

[3.2] RUBENSTEIN, M.F., and R. ROSEN (1968): "Structural Analysis by Matrix Decomposition," *Journal of the Franklin Institute*, Vol. 286, No. 4.

Additional Reading

BENJAMIN, J.R., and C.A. CORNELL (1970): *Probability, Statistics, and Decision for Civil Engineers*, McGraw-Hill Book Company, New York.

HAUGEN, E.B. (1968): *Probabilistic Approaches to Design*, John Wiley & Sons, Inc., New York.

RUBINSTEIN, M.F. (1975): *Patterns of Problem Solving*, Prentice-Hall, Inc., Englewood Cliffs, NJ.

GORDON, G., and I. PRESSMAN (1978): *Quantitative Decision-Making for Business*, Prentice-Hall, Inc., Englewood Cliffs, NJ.

PROBLEMS

3.1   Expand each of the following random functions in a Taylor's series expansion:

(a)   From Example 2-7 consider the beam displacement

$$\Delta = \frac{Pl^3}{3EI}$$

where E, I, and l are deterministic, and the random variable Δ is a function of the random variable P. Expand Δ about the mean value of P, i.e., $\bar{P}$.

(b)   From Problem 2.8 consider the member stress

$$f = \frac{My}{I}$$

where y and I are deterministic functions, and the random variable f is a function of the random variable M. Expand f about the mean value of M, i.e., $\bar{M}$.

(c)   From Example 2-7 consider the support moment

$$M = \frac{Pl}{8} + \frac{\omega l^2}{12}$$

where l is deterministic, and the random variable M is a function of the two random variables P and ω. Expand M about the mean values of P and ω, i.e., $\bar{P}$ and $\bar{\omega}$.

(d)   From Example 2-7 again consider the beam displacement

$$\Delta = \frac{Pl^3}{3EI}$$

where now l and E are deterministic, and the random variable Δ is a function of the two random variables P and I. Expand Δ about the mean values of P and I, i.e., $\bar{P}$ and $\bar{I}$.

3.2   Using the results from Problem 3.1 state in which of the cases one expects a linear statistical analysis to result in no errors due to the Taylor's series truncation.

3.3   From Example 2-7, parts (b) and (c), the two functions, M and Δ, can be related to the two random parameters, P and ω by the equations

$$M = \frac{Pl}{8} + \frac{\omega l^2}{12}$$

$$\Delta = \frac{Pl^3}{192EI} + \frac{\omega l^4}{384EI}$$

(a) Expand the functions M and Δ into a Taylor's series about the mean values of P and ω, i.e., $\bar{P}$ and $\bar{\omega}$.

(b) Calculate the sensitivity matrix, relating changes in M and Δ to changes in P and ω.

3.4   Consider Problem 3.1, part (d). Calculate the sensitivities relating the beam displacement to the random parameters P and I.

3.5   Use a linear statistical analysis to calculate, with your results from Problem 3.1, part (b), the mean and variance of member stress f when y and I are deterministic and M is the random parameter.

3.6   Use a linear statistical analysis to calculate, with your results from Problem 3.1, part (c), the mean and variance of support moment M when l is deterministic and P and ω are random variables.

3.7   In Example 2-7 the equation for the beam displacement is given as

$$\Delta = \frac{Pl^3}{3EI}$$

(a) Let l, E, and I be deterministic. Calculate, using a linear statistical analysis, the mean and variance of Δ when P is a random variable.

(b) Let l and E be deterministic. Calculate, using a linear statistical analysis, the mean and variance of Δ when P and I are random variables.

3.8   A truss structure supports a monorail vehicle which transmits two wheel loads each equal to P to the ground (see Fig. P3-8). The monorail track-support deck is assumed to be rigid, and its weight is negligible compared with the loads P. Assume that the truss is supported against out-of-plane

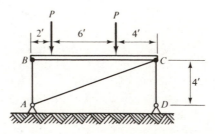

FIGURE P3-8

motion (therefore, consider only a planar truss structure). The wheel loads are assumed to be identical and perfectly correlated (i.e., the correlation coefficient is equal to unity). Denote the mean and standard deviation of P as $\bar{P}$ and σ_p, respectively.

(a) Calculate, using a linear statistical analysis, the mean and standard deviation of the reactions at A and D.

(b) Calculate, using a linear statistical analysis, the mean and standard deviation of the forces in all truss members.

3.9 Consider the truss structure composed on uniform bars as shown in Fig. P3-9. The load acting on the structure P is assumed to be the only random variable, with its mean denoted as $\bar{P}$ and standard deviation denoted as σ_p.

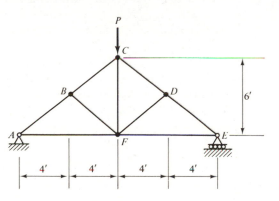

FIGURE P3-9

(a) Calculate, using a linear statistical analysis, the mean and standard deviation of the support reactions at A and E.

(b) Calculate, using a linear statistical analysis, the mean and standard deviation of the force in truss member AF.

3.10 A simple beam supports a uniformly distributed downward load of magnitude ω_0 (Fig. P3-10). The equation for the deflection of the beam (positive being upward) is

$$v(x) = -\left(\frac{\omega_0}{24EI}\right)(L^3 x - 2Lx^3 + x^4)$$

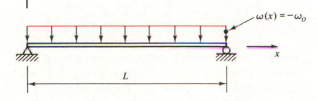

FIGURE P3-10

where

$$E = \text{modulus of elasticity}$$

$$L = \text{length of beam}$$

$$I = \text{moment of inertia}$$

(a) Assume that E, L, and I are deterministic and ω_0 is a random parameter with mean $\bar{\omega}_0$ and standard deviation σ_{ω_0}. Use a linear statistical analysis to calculate the equation for the mean and standard deviation of the beam deflection as a function of x.

(b) Assume E is deterministic and I and ω_0 are independent random parameters. Denote the mean and standard deviation of I as $\bar{I}$ and σ_I, respectively. Use a linear statistical analysis to calculate the equation for the mean and standard deviation of the beam deflection as a function of x.

(c) Explain how your Taylor's series expansion in parts (a) and (b) differ and why your solution in part (a) is exact, whereas your solution in part (b) is only an approximation.

3.11   For the differential element shown in Fig. P3-11 the normal stress and shearing stress for a plane at any angle θ are

$$\sigma_{x'} = \left(\frac{\sigma_x + \sigma_y}{2}\right) + \left(\frac{\sigma_x - \sigma_y}{2}\right)\cos 2\theta + \tau_{xy}\sin 2\theta$$

and

$$\tau_{x'y'} = -\left(\frac{\sigma_x + \sigma_y}{2}\right)\sin 2\theta + \tau_{xy}\cos 2\theta$$

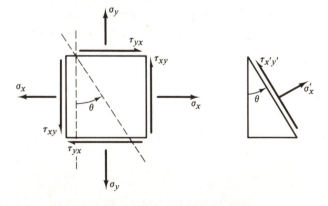

FIGURE P3-11

The positive directions for all stresses are as shown. Let σ_x, σ_y, and τ_{xy} be random parameters with a known covariance matrix.

(a) Calculate the sensitivity matrix relating the random functions $\sigma_{x'}$ and $\tau_{x'y'}$ to the random parameters σ_x, σ_y, and τ_{xy}. Let θ be known.

(b) Let the random parameters be independent random variables with the following properties:

$$\text{mean of } \sigma_x = 3 \text{ ksi}$$
$$\text{standard deviation of } \sigma_x = 0.3 \text{ ksi}$$
$$\text{mean of } \sigma_y = 1 \text{ ksi}$$
$$\text{standard deviation of } \sigma_y = 0.1 \text{ ksi}$$
$$\text{mean of } \tau_{xy} = 2 \text{ ksi}$$
$$\text{standard deviation of } \tau_{xy} = 1 \text{ ksi}$$

Calculate, using a linear statistical analysis, the mean and coefficient of variation of the normal stress and the shearing stress for $\theta = -22.5°$. Are the normal stress and shearing stress independent random variables?

3.12 Consider the same differential element mentioned in Problem 3.11. The principal normal stress is the largest possible stress obtained as the angle θ varies. The equation for the maximum normal stress is

$$\sigma_{\max} = \left(\frac{\sigma_x + \sigma_y}{2}\right) + \sqrt{\left(\frac{\sigma_x - \sigma_y}{2}\right)^2 + \tau_{xy}^2}$$

(a) Use a linear statistical analysis to calculate the mean and standard deviations for this maximum normal stress when σ_x, σ_y, and τ_{xy} are independent random variables with means and standard deviations as given in part (b) of Problem 3.11.

(b) Will σ_x, σ_y, and τ_{xy} always be statistically independent random variables for structural engineering applications? Why? How would your analysis differ if they are not statistically independent?

3.13 A nonlinear statistical analysis utilizes additional terms in the Taylor's series expansion. For example, a second-order nonlinear statistical analysis utilizes the first three terms in the expansion. Consider the equation

$$\Delta = \frac{Pl^3}{3EI}$$

where P, l, and E are deterministic and I is the random parameter. The three-term Taylor's series expansion about the mean value of I, i.e., $\bar{I}$, is

$$\Delta(I) = \frac{Pl^3}{3E\bar{I}} - \left(\frac{Pl^3}{3E\bar{I}^2}\right)(I - \bar{I}) + \frac{1}{2}\left(\frac{2Pl^3}{3E\bar{I}^3}\right)(I - \bar{I})^2$$

Therefore, for a linear statistical analysis the mean would be

$$\bar{\Delta} = \frac{Pl^3}{3E\bar{I}}$$

whereas for a second-order nonlinear statistical analysis the mean would be

$$\bar{\Delta} = \left(\frac{Pl^3}{3E\bar{I}}\right) + \left(\frac{Pl^3}{3E\bar{I}^3}\right) \text{Var}(I)$$

(a) Calculate and compare the two corresponding results for the variance of the displacement Δ.

(b) Is knowledge of the mean and variance of I sufficient information to solve the nonlinear case?

3.14 The equation of motion for a viscously damped single-degree-of-freedom system is

$$m\ddot{x}(t) + c\dot{x}(t) + kx(t) = f(t)$$

where

$$m = \text{mass}$$
$$c = \text{damping}$$
$$k = \text{stiffness}$$
$$f(t) = \text{forcing function}$$
$$x(t) = \text{displacement response}$$
$$\dot{x}(t), \ddot{x}(t) = \text{velocity and acceleration response}$$

Consider m, c, and f to be deterministic and k to be a random variable. Therefore, the displacement is also a random variable and a function of both time and k. A linear Taylor's series expansion of the displacement about the mean value of k, i.e., $\bar{k}$ is

$$x(t, k) = x(t, k = \bar{k}) + \frac{\partial x(t, k)}{\partial k}\bigg|_{k=\bar{k}} (k - \bar{k})$$

The first term is obtained by solving the equation of motion with $k = \bar{k}$, i.e.,

$$m\ddot{x}(t) + c\dot{x}(t) + \bar{k}x(t) = f(t)$$

The $x(t)$ so obtained is $x(t, k = \bar{k})$. The solution for the sensitivity term $\partial x(t, k)/\partial k|_{k=\bar{k}}$ is obtained by noting that

$$\frac{\partial \dot{x}}{\partial k} = \frac{\partial}{\partial k}\left(\frac{dx}{dt}\right) = \frac{d}{dt}\left(\frac{\partial x}{\partial k}\right)$$

and the following:

(1) $m\ddot{x}(t) + c\dot{x}(t) + kx(t) = f(t)$

(2) $\left(\frac{\partial m}{\partial k}\right)\ddot{x}(t) + m\frac{\partial \ddot{x}}{\partial k} + \left(\frac{\partial c}{\partial k}\right)\dot{x}(t) + c\frac{\partial \dot{x}}{\partial k} + \left(\frac{\partial k}{\partial k}\right)x(t) + k\frac{\partial x}{\partial k} = \frac{\partial f}{\partial k}$

(3) $\dfrac{\partial m}{\partial k} = \dfrac{\partial c}{\partial k} = \dfrac{\partial f}{\partial k} = 0$

(4) $m\left(\dfrac{\partial \ddot{x}}{\partial k}\right) + c\left(\dfrac{\partial \dot{x}}{\partial k}\right) + k\left(\dfrac{\partial x}{\partial k}\right) = -x$

(5) $\dfrac{\partial \dot{x}}{\partial k} = \dfrac{d}{dt}\left(\dfrac{\partial x}{\partial k}\right), \qquad \dfrac{\partial \ddot{x}}{\partial k} = \dfrac{d^2}{dt^2}\left(\dfrac{\partial x}{\partial k}\right)$

(6) Defining $z \equiv \dfrac{\partial x}{\partial k}, \qquad \dot{z} = \dfrac{d}{dt}\left(\dfrac{\partial x}{\partial k}\right), \qquad \ddot{z} = \dfrac{d^2}{dt^2}\left(\dfrac{\partial x}{\partial k}\right)$

(7) $m\ddot{z}(t) + c\dot{z}(t) + kz(t) = -x(t)$

Therefore, the solution to (7) for $k = \bar{k}$ and $x(t) = x(t, k = \bar{k})$ results in a solution for $z(t) = \dfrac{\partial x}{\partial k}\Big|_{k=\bar{k}}$. Note that (1) and (7) can be solved by using the same computer program. Therefore, the mean response is

$$\bar{x}(t) = x(t, k = \bar{k})$$

and the variance of the response is

$$\text{Var}\,(x(t)) = \left(\dfrac{\partial x(t, k)}{\partial k}\Big|_{k=\bar{k}}\right)^2 \text{Var}\,(k) = (z(t))^2\,\text{Var}\,(k)$$

(a) Develop the corresponding equations for the mean and variance of the response $x(t)$ if m, k, and $f(t)$ are deterministic and c is a random variable.

(b) Develop the corresponding equations for the mean and variance of the response $x(t)$ if m and $f(t)$ are deterministic and both c and k are random variables.

(c) In parts (a) and (b) the mean and variance is a function of time. Why is this logical?

3.15 Assume that for A-36 structural steel the yield stress is a random variable with mean and standard deviation equal to 47.9 ksi and 3.3 ksi, respectively. Use the first 20 random numbers in column one of Table 3-1 and generate 20 random values for the yield stress, assuming that it has a uniform probability density function.

3.16 In Example 2-2 the sample mean and sample coefficient of variation for the maximum compressive strength of concrete were calculated to be equal to 7.30×10^3 psi and 18.8%, respectively. Generate, using the first 20 random numbers in column one of Table 3-1, 20 random values for the maximum compressive strength, assuming a uniform probability density function with a mean and coefficient of variation equal to 7.30×10^3 psi and 18.8%, respectively.

3.17   Consider the following probability density function:

$$p(x_1) = 0 \qquad\qquad\qquad x_1 < 3.2 \times 10^6 \text{ psi}$$
$$= c_0(x_1 - 3.2 \times 10^6) \qquad 3.2 \times 10^6 \text{ psi} \le x_1 < 3.5 \times 10^6 \text{ psi}$$
$$= c_0(3.8 \times 10^6 - x_1) \qquad 3.5 \times 10^6 \text{ psi} \le x_1 < 3.8 \times 10^6 \text{ psi}$$
$$= 0 \qquad\qquad\qquad x_1 \ge 3.8 \times 10^6 \text{ psi}$$

(a)   Calculate c_0.

(b)   Using the first 20 random numbers in column one of Table 3-1, generate 20 random numbers for the random variable x_1.

3.18   A series of tests are performed on circular reinforced concrete columns. The random variable, denoted R, is

$$R = \left(\frac{\text{test ultimate load}}{\text{design ultimate load}}\right)$$

The design ultimate load is calculated by using the ACI 318-56 code formula,

$$P_u = \text{design ultimate load}$$

$$= \left(\frac{A_s f_y}{\frac{3e'}{d} + 1}\right) + \left[\frac{A_g f'_c}{\frac{9.6De'}{(0.8D + 0.67d)^2} + 1.18}\right]$$

where

A_s = area of tensile reinforcement

f_y = yield stress of reinforcing steel

e' = eccentricity of load

d = diameter of reinforcement circle

A_g = area of concrete section

f'_c = maximum 28 day compressive stress of concrete

D = diameter of column

The sample statistics of R are

$$\text{mean} = 1.028$$
$$\text{coefficient of variation} = 0.125$$

Assume that R has a normal PDF. Generate, using the first 20 random numbers in column one of Table 3-1, 20 random values for R.

3.19   A series of flexure tests are conducted on reinforced concrete beams. The random variable, denoted R, is

$$R = \left(\frac{\text{ultimate test moment}}{\text{ultimate computed moment}}\right)$$

The ultimate computed moment is calculated by using the ACI 318-63 code formula,

$$M_u = \text{ultimate moment}$$

$$= (A_s - A_s')f_y\left(d - \frac{a}{2}\right) + A_s' f_y(d - d')$$

where

A_s = area of tensile reinforcement

A_s' = area of compressive reinforcement

f_y = yield stress of reinforcing steel

d = effective depth

d' = distance from extreme fiber to compressive reinforcement

a = depth of equivalent rectangular stress block

The sample mean and standard deviation of R is 1.113 and 0.083, respectively. Assume that R has a normal PDF and generate, using the first 20 random numbers in column one of Table 3-1, 20 random values for R.

3.20 Repeat Problem 3.16, but now assume that R has a log-normal PDF.

3.21 In Example 2-7 the equation for beam displacement is given as

$$\Delta = \frac{Pl^3}{3EI}$$

Let P, l, and I be deterministic and define a normalized displacement as

$$\Delta' \equiv \frac{\Delta}{Pl^3/3I} = \frac{1}{E}$$

Consider E to be a uniformly distributed random variable with mean and coefficient of variation equal to 3.52×10^6 psi and 4.2%, respectively.

(a) Use the first 20 random numbers in column one of Table 3-1 to generate 20 random values for E.

(b) Use the Monte Carlo method to calculate the sample mean and sample variance for Δ'.

(c) Calculate, using a linear statistical analysis, the mean and standard deviation of Δ'.

(d) Discuss why your answers to parts (b) and (c) are different.

(e) Assuming that instead of using just 20 samples in your Monte Carlo analysis you used an infinite number of samples, would your answers to parts (b) and (c) still differ? If so, why?

3.22 A reinforced-concrete beam with the section shown in Fig. P3-22 is subjected to a bending moment which produces tension in the steel bars. The steel

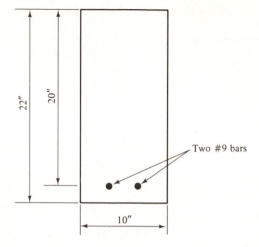

reinforcement consists of two #9 bars. (A #9 bar is $1\frac{1}{8}$ in. in diameter and has a cross-sectional area of 1 sq in.). Plane sections are assumed to remain plane, and the strains are assumed to vary linearly from the neutral axis. Assume that the concrete cannot resist tension.

(a) Assume that the modulus of elasticities of the concrete and steel are deterministic with values equal to 2×10^6 psi and 30×10^6 psi, respectively. The bending moment is considered to be a random variable with mean and standard deviations equal to 50 kip-ft and 10 kip-ft, respectively. Calculate, using a linear statistical analysis, the mean and standard deviation of the maximum concrete compressive stress and the maximum steel tensile stress. Also calculate the covariance between these two stresses.

(b) Calculate the mean and standard deviation of the maximum concrete compressive stress, using a Monte Carlo analysis, with the modulus of elasticity of the steel and concrete *also* considered to be random variables with means as noted above (i.e., 2×10^6 psi and 30×10^6 psi) and with the coefficient of variations for the modulus of elasticities of concrete and steel equal to 10 and 5%, respectively. Assume that the applied bending moment and the modulus of elasticities are all independent uniform random variables. To generate the Monte Carlo numbers use the first 20 random numbers in column one of Table 3-1 for the moment, the first 20 random numbers in column two of Table 3-1 for the modulus of elasticity of the concrete, and the first 20 random numbers in column three of Table 3-1 for the modulus of elasticity of the steel.

3.23 A 4 in. wide by 6 in. deep (actual size) uniform wood beam is used to support a uniformly distributed load of ω pounds per inch on a simple span of 10 ft. The applied load acts in a plane making an angle θ with the vertical. The maximum moment at the midspan can be resolved into components acting

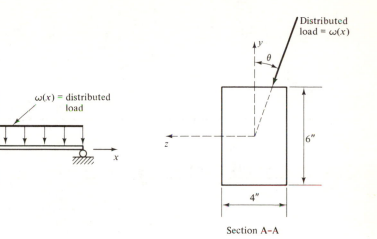

FIGURE P3-23

around the *y-y* and *z-z* axes. These moments are

$$M_{zz} = \left(\frac{\omega L^2}{8}\right)\cos\theta$$

$$M_{yy} = \left(\frac{\omega L^2}{8}\right)\sin\theta$$

The distributed load acts along a line through the centroid of the cross section. Consider the geometric properties of the beam to be deterministic.

(a) If the angle θ is deterministic and equal to 30°, calculate, using a linear statistical analysis, the mean and standard deviations of the maximum tensile stress on the cross section. Assume the uniform load to be a random variable with a mean and standard deviation of 10 lb/ft and 2 lb/ft, respectively.

(b) Now, consider the angle θ to *also* be a random variable, independent of the distributed load ω, with a mean of 30° and a standard deviation equal to 5°. Assume that both ω and θ have uniform PDF's. Calculate the mean and standard deviations of the maximum tensile stress on the cross section. Use, to generate your random Monte Carlo numbers, the first 20 random numbers in column one of Table 3-1 for ω and the first 20 random numbers in column two of Table 3-1 for θ.

Structural Safety

4-1 INTRODUCTION

In Chapter 2 we discussed the quantification of uncertainty by using probability and statistics. Therefore, parameters describing structural resistance can now be visualized by the reader as random variables with associated probability density functions. Chapter 3 indicates how one can describe the response given a probabilistic description of the structural parameters. In the next chapter structural loads are discussed and special consideration is given to the statistical characterization of these loads. Engineering failures do occur. Often these failures result from a combination of large values of loading and low values of resistant capacity. In this chapter we discuss the concept of structural safety by introducing the term *failure* and calculating the probability of failure.

Dictionaries define failure as being deficient or negligent in an obligation, duty, or expectation. Therefore, using this definition, failure is a function of one's perspective or one's obligation or duty.

4-2 PROBABILITY OF FAILURE: TWO SPECIAL CASES

To initiate the discussion of probability of failure let us consider a special case. A single structural component (see Fig. 4-1) is acted upon by a random load. The mean and standard deviation of the load are assumed known and denoted as $\bar{P}$ and σ_p, respectively. The load induces a stress, and if the beam cross-sectional area is considered to be deterministic the stress is a random variable S with mean $\bar{S} = \bar{P}/A$ and

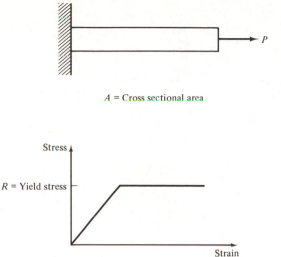

FIGURE 4-1. Hypothetical example.

standard deviation $\sigma_S = \sigma_p/A$. Assume that the stress-strain properties of the material are as shown in Figure 4-1. The yield stress of the material is considered to be a random variable R with known mean and standard deviation (denoted as $\bar{R}$ and σ_R). If failure is *defined* to exist when the load-induced stress equals or exceeds the material yield stress, then failure exists when

$$S \geq R \tag{4-1}$$

If one defines a new random variable F as

$$F \equiv R - S \tag{4-2}$$

failure therefore occurs when F is less than or equal to zero.

Consider first the special case where both *R and S have independent normal PDF's.* That is,

$$p(s) = \frac{1}{\sigma_S \sqrt{2\pi}} \exp\left\{-\frac{1}{2}\left(\frac{s - \bar{S}}{\sigma_S}\right)^2\right\} \tag{4-3}$$

and

$$p(r) = \frac{1}{\sigma_R \sqrt{2\pi}} \exp\left\{-\frac{1}{2}\left(\frac{r - \bar{R}}{\sigma_R}\right)^2\right\} \tag{4-4}$$

The random variable F has a normal probability density function because it is a linear combination of two normally distributed random variables. It directly follows (see Example 2-9) that

$$\bar{F} = \bar{R} - \bar{S} \tag{4-5}$$

and

$$\sigma_F^2 = \sigma_R^2 + \sigma_S^2 \tag{4-6}$$

Therefore, the PDF of F is

$$p(f) = \frac{1}{\sigma_F \sqrt{2\pi}} \exp \left\{ -\frac{1}{2} \left(\frac{f - \bar{F}}{\sigma_F} \right)^2 \right\} \tag{4-7}$$

and is shown in Figure 4-2. Since failure occurs when F is less than or equal to zero, the *probability of failure* P_f is

$$P_f = \Pr [F \leq 0] = \int_{-\infty}^{0} p(f) \, df \tag{4-8}$$

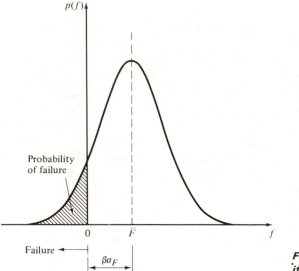

FIGURE 4-2. Failure probability.

Figure 4-3 shows the calculated probability of failure for several combinations of coefficients of variations of R and S (i.e., $\rho_R = (\sigma_R / \bar{R})$ and $\rho_S = (\sigma_S / \bar{S})$) and several *central safety factor* values where

$$\text{central safety factor} \equiv C_0 \equiv \frac{\bar{R}}{\bar{S}} \tag{4-9}$$

It is apparent from Figure 4-2 that $F = 0$ occurs at $\bar{F}$ minus β standard deviations. Therefore, the β value is directly related to the probability of failure. It follows that knowing $\bar{F}$ and σ_F and setting $F = 0$, i.e.,

$$F = 0 = \bar{F} - \beta \sigma_F$$

results in a value of β equal to

$$\beta = \frac{\bar{R} - \bar{S}}{\sqrt{\sigma_R^2 + \sigma_S^2}} = \frac{\bar{F}}{\sigma_F} \tag{4-10}$$

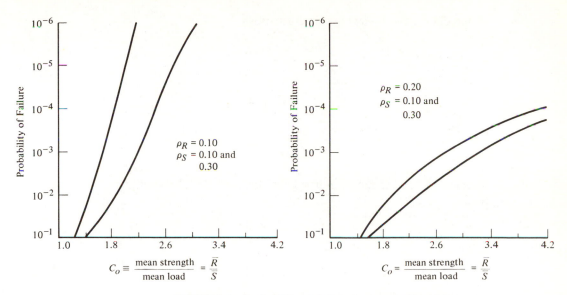

FIGURE 4-3. Probability of failure for normal R and S.

β is called the *reliability index* or safety index. Equation (4-7) can be transformed into the standardized normal PDF (see Section 2-9):

$$p(u) = \frac{1}{\sqrt{2\pi}} \exp\left\{-\frac{1}{2}u^2\right\} \tag{4-11}$$

if

$$U \equiv \frac{F - \bar{F}}{\sigma_F} \tag{4-12}$$

The probability of failure is now obtained when $F \leq 0$ (i.e., $u = -(\bar{F}/\sigma_F) = -\beta$), and is

$$P_f = \Pr[F \leq 0] = \int_{-\infty}^{-\beta} p(u)\, du \tag{4-13}$$

Table 2-4 is used to obtain P_f for a given β or, alternatively, a value of β for a given P_f. Although the specification of acceptable P_f or β values is beyond the scope of this book, it is instructive to note that representative values which are commonly accepted are

strength failure: $\qquad P_f = 10^{-4}, \beta \approx 3.5$

serviceability failure: $\quad P_f = 10^{-2}, \beta \approx 2.0$

Consider the second special case, where *R and S have independent log-normal PDF's*. Taking the natural logarithm of the ratio of *R* divided by *S*, it follows that

$$\ln\left(\frac{R}{S}\right) = \ln R - \ln S \tag{4-14}$$

If one defines the following transformations,

$$X \equiv \ln R$$

$$Y \equiv \ln S$$

$$Z \equiv \ln \left(\frac{R}{S}\right)$$

then Eq. (4-14) becomes

$$Z = X - Y \tag{4-15}$$

where X, Y, and Z are all normally distributed. It follows directly from Eq. (4-15) that

$$\bar{Z} = \bar{X} - \bar{Y} \tag{4-16}$$

and

$$\sigma_Z^2 = \sigma_X^2 + \sigma_Y^2 \tag{4-17}$$

and

$$p(z) = \frac{1}{\sigma_Z \sqrt{2\pi}} \exp\left\{-\frac{1}{2}\left(\frac{z - \bar{Z}}{\sigma_Z}\right)^2\right\} \tag{4-18}$$

It now remains that we find $\bar{X}$, $\bar{Y}$, σ_X^2 and σ_Y^2 in terms of $\bar{R}$, $\bar{S}$, σ_R^2 and σ_S^2.

Recall from Section 2-9 that the mean and variance of the transformed variable (i.e., X) can be expressed in terms of the mean and variance of the log-normal random variable.

It follows from the power series expansion of $\ln(1 + \epsilon)$ that this is equal to ϵ for small values of ϵ. In the present case, ϵ is equal to the coefficient of variation squared, and therefore this approximation is acceptable. Thus, it is possible to use the following approximate relationships:

$$\bar{X} = \ln \bar{R} \tag{4-19}$$

and

$$\sigma_X^2 = \rho_R^2 \tag{4-20}$$

where

$$\rho_R = \text{coefficient of variation of } R$$

Therefore, Eqs. (4-16) and (4-17) become

$$\bar{Z} = \ln \bar{R} - \ln \bar{S} \tag{4-21}$$

and

$$\sigma_Z^2 = \rho_R^2 + \rho_S^2 \tag{4-22}$$

Recall that failure occurs when

$$R - S < 0$$

or, alternatively, when

$$\frac{R}{S} < 1$$

Taking the natural logarithm of both sides, we obtain

$$\ln \left(\frac{R}{S}\right) = \ln R - \ln S \leq 0$$

Therefore, failure exists when Z as defined by Eq. (4-15) is equal to or less than zero.

The normal PDF in Eq. (4-18) can be transformed into the standardized normal PDF by defining

$$U^* \equiv \frac{Z - \bar{Z}}{\sigma_Z} \tag{4-23}$$

It directly follows that

$$p(u^*) = \frac{1}{\sqrt{2\pi}} \exp\left\{-\frac{1}{2}(u^*)^2\right\} \tag{4-24}$$

The probability of failure is obtained when $Z \leq 0$ (i.e., $u^* = -(\bar{Z}/\sigma_Z)$), and is

$$P_f = \Pr[Z \leq 0] = \int_{-\infty}^{-\beta^*} p(u^*)\, du^* \tag{4-25}$$

where

$$\beta^* = \frac{\ln \bar{R} - \ln \bar{S}}{\sqrt{\rho_R^2 + \rho_S^2}} \tag{4-26}$$

Equations (4-13) and (4-25) are very similar in form. In the former equation, R and S are assumed to be normally distributed and β is evaluated by using Eq. (4-10). The latter equation is for log-normally distributed R and S random variables, and β^* is evaluated by using Eq. (4-26). In both cases, Table 2-4 is used to calculate the probability of failure.

Example 4-1

The yield stress mean and standard deviation of a steel member are 47.9 ksi and 3.3 ksi, respectively. Assume that the member is loaded such that the induced stress has a mean of 36.0 ksi and a standard deviation of 7.2 ksi. Calculate the probability of failure, assuming that the material yield stress and load-induced stress are independent and both normally distributed.

Using the above it follows directly that

$$\bar{R} = 47.9 \text{ ksi}$$

$$\sigma_R = 3.3 \text{ ksi}$$

$$\bar{S} = 36.0 \text{ ksi}$$

$$\sigma_S = 7.2 \text{ ksi}$$

Using Eq. (4-9) the central safety factor is

$$C_0 \equiv \frac{\bar{R}}{\bar{S}} = \frac{47.9}{36.0} = 1.33$$

Note that this factor of safety does not depend upon the uncertainty in either the loading-induced stress or the material yield stress.

The mean and standard deviation of F follow from Eqs. (4-5) and (4-6) and are

$$\bar{F} = \bar{R} - \bar{S} = 11.9 \text{ ksi}$$

and

$$\sigma_F = \sqrt{\sigma_R^2 + \sigma_S^2} = \sqrt{10.89 + 51.84} = 7.92 \text{ ksi}$$

Therefore, the value of β follows from Eq. (4-10) and is equal to 1.5. The probability of failure is (see Table 2-4):

$$P_f = \int_{-\infty}^{-\beta} p(u)\, du = \int_{-\infty}^{-1.50} p(u)\, du = 1 - \int_{-\infty}^{+1.5} p(u)\, du$$

$$= 1.000 - 0.933 = 0.067, \text{ or } 6.7\%$$

Note that *if* the loading were known with more accuracy, then P_f would decrease but the central factor of safety would remain unchanged. For example, if the load-induced stress had a coefficient of variation of 10% instead of 20%, then

$$\bar{F} = 11.9 \text{ ksi}$$

$$\sigma_F = \sqrt{10.89 + 12.96} = 4.88 \text{ ksi}$$

$$\beta = 2.439$$

$$P_f = 1.000 - 0.993 = 0.007, \text{ or } 0.7\%$$

Example 4-2

In Example 3-1 we calculated the mean and standard deviation of the load-induced bending moment at a distance of 6 ft from the free end of a cantilever beam. These values are

$$\bar{M} = \text{mean load-induced moment} = -23.1 \text{ kip-ft}$$

$$\sigma_M = \text{standard deviation of the load-induced moment} = 3.6 \text{ kip-ft}$$

Assume that the load-induced moment has a normal PDF. The bending moment capacity of the beam is now assumed to be a random variable with a normal PDF. The load-induced bending moment and the bending moment capacity are assumed to be independent random variables.

(a) Assume that the beam cross section has been designed such that the central safety factor relating load-induced moment and bending capacity moment is 1.5. Also assume that the coefficient of variation of the bending moment capacity is 15%. Calculate the probability of failure if failure is defined to occur whenever the load-induced moment exceeds the bending capacity moment.

(b) If the probability of failure is required to be less than $\frac{1}{2}\%$, what is the minimum required mean value of bending moment capacity? Assume that the coefficient of variation of the bending moment capacity remains equal to 15%.

Part (a) solution:

The central safety factor is equal to

$$C_0 = \frac{\bar{M}_c}{\bar{M}} = \frac{\text{mean bending moment capacity}}{\text{mean load-induced moment}}$$

Therefore, if $C_0 = 1.5$ it follows that

$$\bar{M}_c = 1.5(23.1) = 34.7 \text{ kip-ft}$$

It follows from the problem statement that the standard deviation of the bending moment capacity (σ_{M_c}) is

$$\sigma_{M_c} = 0.15(34.7) = 5.2 \text{ kip-ft}$$

Failure has been defined to occur when the moment capacity is equal to or less than the load-induced moment. Therefore, if

$$F = M_c - M$$

then

$$F \leq 0 \quad \text{Failure}$$
$$F > 0 \quad \text{Safe}$$

The mean and standard deviation of F follow directly and are equal to

$$\bar{F} = \bar{M}_c - \bar{M} = 34.7 - 23.1 = 11.6 \text{ kip-ft}$$

and

$$\sigma_F = \sqrt{(3.6)^2 + (5.2)^2} = 6.32 \text{ kip-ft}$$

The probability of failure is

$$P_f = \int_{-\infty}^{-\beta} p(u) \, du = \int_{-\infty}^{-1.84} p(u) \, du = 0.033$$

where

$$\beta = \frac{\bar{F}}{\sigma_F} = \frac{11.6}{6.32} = 1.84$$

Part (b) solution:

This solution represents a design problem where one is required to determine the mean value of the bending moment capacity given the coefficient of variation of the bending moment capacity and the acceptable probability of failure.

The probability of failure is

$$P_f = \int_{-\infty}^{-\beta} p(u) \, du$$

Therefore, given $P_f = 0.005$ it follows that $\beta = 2.71$

Recalling the definition of β it follows that

$$\beta = \frac{\bar{F}}{\sigma_F} = \frac{\bar{M}_c - \bar{M}}{\sqrt{\sigma_{M_c}^2 + \sigma_M^2}} = \frac{\bar{M}_c - 23.1}{\sqrt{(0.15\bar{M}_c)^2 + (3.6)^2}} = 2.71$$

and finally

$$\bar{M}_c = 43.2 \text{ kip-ft}$$

Therefore, if $\bar{M}_c$ is equal to or greater than 43.2 kip-ft, then the stated design objective is satisfied.

Example 4-3

Recall from Example 3-9 that the structure was assumed to collapse when the maximum ground acceleration (denoted a) exceeded the acceleration level a_3. In that example a_3 was assumed to be deterministic and equal to 125 cm/sec. Now assume that this collapse level acceleration is a random variable with a normal PDF. That is,

$$p(a_3) = \frac{1}{\sigma_{a_3}\sqrt{2\pi}} \exp\left\{-\frac{1}{2}\left(\frac{a_3 - \bar{a}_3}{\sigma_{a_3}}\right)^2\right\}$$

where

$$\bar{a}_3 = \text{mean collapse level acceleration}$$

$$\sigma_{a_3} = \text{standard deviation of collapse level acceleration}$$

Failure of the structure is now defined to be collapse. Therefore, failure occurs when the maximum ground acceleration exceeds the collapse level acceleration. If one defines

$$F = a_3 - a$$

then when F is less than or equal to zero failure occurs. For this example assume that a and a_3 are independent and normally distributed random variables.

  (a) Calculate the central safety factor against collapse.
  (b) Calculate the probability of failure.
  (c) If σ_{a_3} is zero the collapse level acceleration is a deterministic variable. Compare the β for this case with the β for a random variable collapse level acceleration.

─────────────────────────────

Part (a) solution:

The central safety factor is equal to

$$C_0 = \frac{\bar{a}_3}{\bar{a}}$$

Part (b) solution:

The mean and standard deviation of F are

$$\bar{F} = \bar{a}_3 - \bar{a}$$

and

$$\sigma_F = \sqrt{\sigma_{a_3}^2 + \sigma_a^2}$$

Therefore, the probability of failure is

$$P_f = \int_{-\infty}^{0} \frac{1}{\sqrt{\sigma_{a_3}^2 + \sigma_a^2}\sqrt{2\pi}} \exp\left\{-\frac{1}{2}\left(\frac{f - \bar{a}_3 + \bar{a}}{\sqrt{\sigma_{a_3}^2 + \sigma_a^2}}\right)^2\right\} df$$

or

$$P_f = \int_{-\infty}^{-\beta} p(u)\, du$$

where

$$\beta = \frac{\bar{F}}{\sigma_F} = \frac{\bar{a}_3 - \bar{a}}{\sqrt{\sigma_{a_3}^2 + \sigma_a^2}}$$

Part (c) solution:

If the collapse level acceleration is deterministic, then

$$\sigma_{a_3} = 0$$

and

$$\beta = \left(\frac{\bar{a}_3 - \bar{a}}{\sigma_a} \right)$$

Therefore, the ratio of the deterministic and random β values is

$$\beta \text{ ratio} = \left(\frac{\sqrt{\sigma_{a_3} + \sigma_a^2}}{\sigma_a} \right)$$

$$= \sqrt{1 + \left(\frac{\sigma_{a_3}}{\sigma_a} \right)^2}$$

It is clear that the ratio is always greater than one. Therefore, a larger β value for a deterministic collapse level acceleration results in a *lower* calculated probability of failure.

Example 4-4

Consider a truss member of area A subjected to an axial load P. The load-induced stress is

$$f = \frac{P}{A}$$

where P is a normal random variable and A is deterministic. Assume that the material yield stress (denoted as f_y) is a normal random variable. Assume that P and f_y are independent random variables. The statistics of the random variables are as follows:

$$\bar{P} = \text{mean axial load} = 200 \text{ kips}$$

$$\sigma_P = \text{standard deviation of axial load} = 40 \text{ kips}$$

$$\bar{f}_y = \text{mean yield stress} = 47.9 \text{ ksi}$$

$$\sigma_{f_y} = \text{standard deviation of yield stress} = 3.3 \text{ ksi}$$

(a) If the member is to be designed such that the central safety factor is 1.2, calculate the required truss member area.

(b) Define a deterministic design load to be

$$P_D = \bar{P} + \alpha_{DP}\sigma_P$$

and find the value of the constant α_{DP} such that the probability that the load is greater than P_D is 0.10.

(c) If P_D is written $P_D = \phi_L\bar{P}$, what is the value of ϕ_L for part (b)? The term ϕ_L is called the *load factor*.

(d) Define a deterministic design yield stress level to be

$$f_D = \bar{f}_y - \alpha_{DS}\sigma_{f_y}$$

and find the value of the constant α_{DS} such that the probability that the yield stress is less than f_D is 0.10.

(e) If f_D is written $f_D = \phi_S\bar{f}_y$, what is the value of ϕ_S for part (d)? The term ϕ_S is called the *capacity reduction factor*.

(f) Determine the required truss member area such that $f_D = P_D/A$.

(g) What is the probability of failure if failure is defined to be a load-induced stress equal to or greater than a material yield stress when the area of the truss member is that determined (1) in part (a), (2) in part (f)?

Part (a) solution:

The central safety factor is

$$C_0 = 1.2 = \frac{\bar{f}_y}{\bar{P}/A}$$

Therefore,

$$A = 1.2\frac{\bar{P}}{\bar{f}_y} = 1.2\left(\frac{200}{47.9}\right) = 5.01 \text{ in.}^2$$

Part (b) solution:

The probability that the load is equal to or less than P_D is 0.90, and therefore

$$\Pr[P \le P_D] = 0.90 = \int_{-\infty}^{P_D} \frac{1}{40\sqrt{2\pi}}\exp\left\{-\frac{1}{2}\left(\frac{P-200}{40}\right)^2\right\} dP$$

Using the standardized normal distribution table from Chapter 2 it follows that

$$P_D = 251.3 \text{ kips}$$

Therefore,

$$P_D = \bar{P} + \alpha_{DP}\sigma_P = 200 + \alpha_{DP}(40) = 251.3$$

and

$$\alpha_{DP} = 1.28$$

Part (c) solution:

$$P_D = 251.3 = \phi_L \bar{P} = \phi_L(200)$$

Therefore,

$$\phi_L = \text{load factor} = 1.26$$

Part (d) solution:

The probability that the material yield stress is less than or equal to f_D is 0.10, and therefore

$$\Pr[f_y \leq f_D] = 0.10 = \int_{-\infty}^{f_D} \frac{1}{3.3\sqrt{2\pi}} \exp\left\{-\frac{1}{2}\left(\frac{f_y - 47.9}{3.3}\right)^2\right\} df_y$$

Using the standardized normal distribution table in Chapter 2 it follows that

$$f_D = 43.7 \text{ ksi}$$

Therefore,

$$f_D = \bar{f}_y - \alpha_{DS}\sigma_{f_v} = 47.9 - \alpha_{DS}(3.3) = 43.7$$

and

$$\alpha_{DS} = 1.28$$

Part (e) solution:

$$f_D = 43.7 = \phi_S \bar{f}_y = \phi_S(47.9)$$

Therefore,

$$\phi_S = \text{capacity reduction factor} = 0.91$$

Part (f) solution:

The required truss member area is

$$A = \frac{P_D}{f_D} = \frac{251.3}{43.7} = 5.75 \text{ in.}^2$$

Part (g) solution:

$$F = f_y - \frac{P}{A}$$

and

$$\bar{F} = \bar{f}_y - \frac{\bar{P}}{A}$$

$$\sigma_F^2 = \sigma_{f_v}^2 + \frac{\sigma_P^2}{A^2}$$

The probability of failures P_f is

$$P_f = \int_{-\infty}^{0} \frac{1}{\sigma_F \sqrt{2\pi}} \exp\left\{-\frac{1}{2}\left(\frac{F - \bar{F}}{\sigma_F}\right)^2\right\} dF$$

Therefore, it follows that if

$$A = 5.01 \text{ in.}^2$$
$$\bar{F} = 7.98$$
$$\sigma_F = 8.64$$

and

$$P_f = 0.178$$

Similarly, if

$$A = 5.75 \text{ in.}^2$$
$$\bar{F} = 13.12$$
$$\sigma_F = 7.70$$

and

$$P_f = 0.044$$

Note that the two alternative deterministic design approaches result in two different member areas and hence two different probabilities of failure. The second approach directly incorporates the uncertainty in the loading and strength of the system.

Example 4-5

Repeat Example 4-1, but now assume that the material yield stress and load-induced stress have independent log-normal PDF's.

In this case, one must calculate β^* instead of β. Using Eq. (4-26) it follows that

$$\beta^* = \frac{\ln \bar{R} - \ln \bar{S}}{\sqrt{\rho_R^2 + \rho_S^2}}$$

where

$$\ln \bar{R} = \ln(47.9) = 3.869$$
$$\ln \bar{S} = \ln(36.0) = 3.584$$
$$\rho_R = \frac{3.3}{47.9} = 0.069$$
$$\rho_S = \frac{7.2}{36.0} = 0.20$$

and finally

$$\beta^* = \frac{3.869 - 3.584}{\sqrt{0.0400 + 0.00476}} = 1.347$$

From Table 2-4 the probability of failure is

$$P_f = \int_{-\infty}^{-1.347} p(u^*) \, du^* = 1 - \int_{-\infty}^{+1.347} p(u^*) \, du^*$$

$$= 1 - 0.911 = 0.088, \text{ or } 8.8\%$$

The relative importance of the type of PDF and the quantification of the numerical values of the statistical parameters is something the reader should now reflect on.

4-3 PROBABILITY OF FAILURE

In the previous section we discussed the calculation of the probability of failure for two special types of PDF for R and S. In a more general discussion of probability of failure, one must not have such a restriction on the mathematical form of the PDF.

What is *failure*? Failure is what the structural engineer defines it to... nothing else. For example, if the stress induced by an earthquake exceeds the yield stress of the material, it could be called failure. Alternatively, if the stress exceeds the ultimate stress of the material, it could be called failure. Failure can also be related to structure serviceability. For example, building interstory displacement greater than $\frac{1}{2}$ in. could be called failure. Therefore, it is fundamentally important to realize that the structural engineer defines failure, and that the examples are virtually unlimited.

Denote the load-induced response (e.g., stress) by the symbol S, where S is understood to be a random variable. The PDF is assumed to be known and is denoted $p_S(x)$. The symbol R is used to denote the structure's resistance (e.g., yield stress). It is a random variable with a PDF denoted by $p_R(x)$. Figure 4-4 shows the two PDF's vs. the response variable (e.g., stress).

Failure exists, as noted in Eq. (4-1), when

$$S > R$$

The probability that S will be equal to or greater than a value denoted x is obtained by evaluating the probability distribution function of S at x, i.e., $P_S(x)$, and subtracting it from unity; i.e.,

$$\text{Pr}\,[S \geq x] = 1 - P_S(x) \tag{4-27}$$

The probability that, for small dx, R will be between x and $x + dx$ is $p_R(x)\,dx$. Therefore, for all possible values of x it follows that failure exists if R is between x and

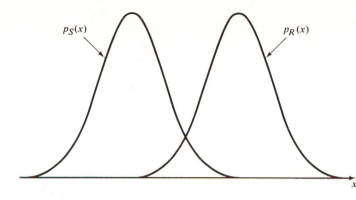

FIGURE 4-4. Response and resistance probability den– sity functions.

$x + dx$ and S is greater than or equal to x. The probability of failure P_f is therefore

$$P_f = \int_{-\infty}^{+\infty} [1 - P_S(x)]p_R(x)\,dx \tag{4-28}$$

The probability of safe behavior (denoted as P_s) is one minus the probability of failure; i.e.,

$$P_s = 1 - P_f \tag{4-29}$$

The *probability of safe behavior* is referred to as *reliability*.

Equation (4-28) can be numerically, if not close form, evaluated to obtain the probability of failure. This equation can be used for general PDF's. Figures 4-5 and

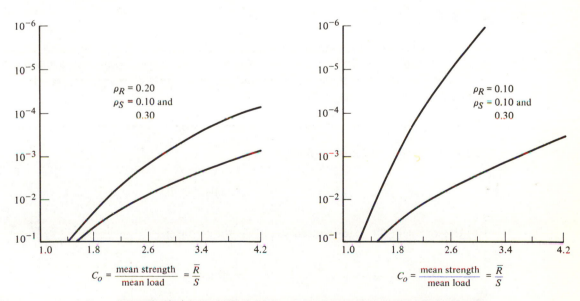

FIGURE 4-5. Probability of failure: R-normal, S-type II largest extreme.

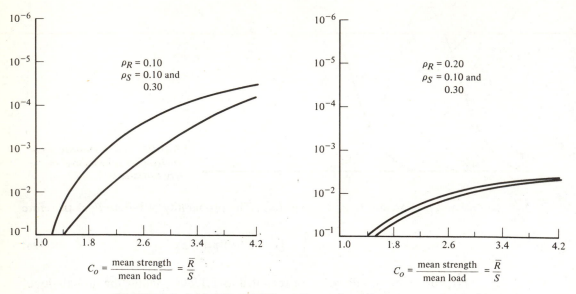

FIGURE 4-6. Probability of failure: R-type I smallest extreme, S-type I largest extreme.

4-6 show the calculated probability of failures for two different combinations of R and S probability density functions.

Example 4-6

Consider a structural member loaded such that it has a PDF of induced stress $p_s(x)$ as shown in Fig. Ex. 4-6. Assume that the material yield stress has a uniform PDF as shown in the same figure. Calculate the probability of failure and reliability of this structural member.

The probability of failure is defined as

$$P_f = \int_{-\infty}^{+\infty} [1 - P_S(x)] p_R(x) \, dx$$

For this problem the load-induced stress has the uniform PDF given by

$$p_s = \left(\frac{1}{37 - 30}\right) = \frac{1}{7} \qquad 30 \text{ ksi} \le x \le 37 \text{ ksi}$$

$$= 0 \qquad\qquad\qquad \text{otherwise}$$

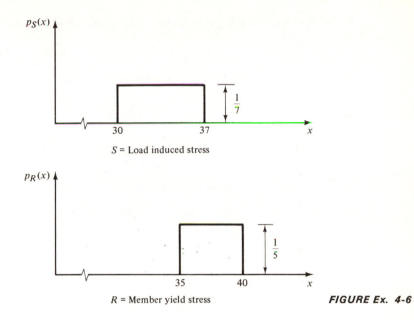

FIGURE Ex. 4-6

The corresponding probability distribution function is

$$P_S(x) = \int_{-\infty}^{x} p_s(x)\, dx$$

$$= 0 \qquad\qquad x < 30 \text{ ksi}$$
$$= \tfrac{1}{7}(x - 30) \qquad 30 \text{ ksi} \le x \le 37 \text{ ksi}$$
$$= 1 \qquad\qquad x > 37 \text{ ksi}$$

The PDF of material yield stress is assumed to be

$$p_R(x) = \left(\frac{1}{40 - 35}\right) = \frac{1}{5} \qquad 35 \text{ ksi} \le x \le 40 \text{ ksi}$$
$$= 0 \qquad\qquad\qquad \text{otherwise}$$

The integral for the probability of failure is therefore separated into five regions as follows:

$$P_f = \int_{-\infty}^{30} [1 - 0](0)\, dx + \int_{30}^{35}\left[1 - \left(\frac{x - 30}{7}\right)\right](0)\, dx$$

$$+ \int_{35}^{37}\left[1 - \left(\frac{x - 30}{7}\right)\right]\left(\frac{1}{5}\right) dx + \int_{37}^{40} [1 - 1]\left(\frac{1}{5}\right) dx + \int_{40}^{+\infty} [1 - 1](0)\, dx$$

$$\therefore \quad P_f = \int_{35}^{37}\left[1 - \left(\frac{x - 30}{7}\right)\right]\left(\frac{1}{5}\right) dx = 0.057, \text{ or } 5.7\%$$

Recall that the reliability is one minus the probability of failure and therefore is

$$P_s = 1 - P_f = 1.000 - 0.057 = 0.942, \text{ or } 94.2\%$$

Example 4-7

Repeat Example 4-1, but now assume that the material yield stress and load-induced stress have independent uniform PDF's.

The mean and variance of the steel yield stress are

$$\bar{R} = 47.9 \text{ ksi} = \left(\frac{a+b}{2}\right)$$

and

$$\sigma_R^2 = (3.3 \text{ ksi})^2 = \frac{(b-a)^2}{12}$$

therefore,

$$a = 42.18 \text{ ksi}$$
$$b = 53.62 \text{ ksi}$$

Similarly, for the load-induced stress it follows that

$$\bar{S} = 36.0 \text{ ksi} = \frac{(c+d)}{2}$$

and

$$\sigma_S^2 = (7.2 \text{ ksi})^2 = \frac{(d-c)^2}{12}$$

Therefore,

$$c = 23.53 \text{ ksi}$$
$$d = 48.47 \text{ ksi}$$

The probability of failure is defined as

$$P_f = \int_{-\infty}^{+\infty} [1 - P_S(x)] p_R(x) \, dx$$

where

$$
\begin{aligned}
p_R(x) &= 0 & x &< a \\
&= 0.087413 & a &\leq x \leq b \\
&= 0 & x &> b
\end{aligned}
$$

and

$$P_S(x) = \int_{-\infty}^{x} P_s(x)\, dx$$

$$\begin{aligned} &= 0 & x < c \\ &= 0.040096(x - 23.53) & c \le x \le d \\ &= 0 & x > d \end{aligned}$$

It therefore follows that

$$P_f = 0.0693$$

This probability of failure compares with 6.7% for the normal case of Example 4-1 and 8.8% for the log-normal case of Example 4-4.

Example 4-8

Consider the material yield stress f_y to be a random variable with a gamma PDF given by

$$p_{f_y}(f_y) = \frac{1}{a\Gamma(b)}\left(\frac{f_y}{a}\right)^{b-1} \exp\left\{-\frac{f_y}{a}\right\} \qquad \begin{matrix} f_y > 0 \\ a, b > 0 \end{matrix}$$

where a and b are established by using the mean and standard deviation of the load-induced stress. That is,

$$\bar{f}_y = ab$$

and

$$\sigma_{f_y}^2 = a^2 b$$

Assume that the load-induced stress f has an exponential PDF given by

$$p_f(f) = \frac{1}{a_1} \exp\left\{-\left(\frac{f - c_1}{a_1}\right)\right\} \qquad \begin{matrix} f > c_1 \\ a_1 > 0 \end{matrix}$$

where in this case a_1 and c_1 are determined from the mean and variance of the yield stress; i.e.,

$$\bar{f} = c_1 + a_1$$

and

$$\sigma_f^2 = a_1^2$$

Failure is defined to exist when the load-induced stress is equal to or greater than the material yield stress; i.e.,

$$f \ge f_y \quad \text{(failure)}$$

Explain how the probability of failure is calculated.

The probability of failure is

$$P_f = \int_{-\infty}^{+\infty} [1 - P_f(x)] p_{f_v}(x)\, dx$$

where

$$P_{f_v}(x) = \frac{1}{a\Gamma(b)}\left(\frac{x}{a}\right)^{b-1} \exp\left\{-\frac{x}{a}\right\} \qquad x > 0$$

and

$$P_f(x) = \int_{-\infty}^{x} \frac{1}{a_1} \exp\left\{-\left(\frac{f - c_1}{a_1}\right)\right\} df$$

$$= 0$$

$$= \exp\left\{-\left(\frac{x - c_1}{a_1}\right)\right\} - 1 \qquad \begin{aligned} x &< c_1 \\ x &> c_1 \end{aligned}$$

Therefore, the probability of failure is

$$P_f = \int_{c_1}^{\infty} \left[-\exp\left\{-\left(\frac{x - c_1}{a_1}\right)\right\} \right] \frac{1}{a\Gamma(b)}\left(\frac{x}{a}\right)^{b-1} \exp\left\{-\frac{x}{a}\right\} dx$$

Note that this integral is best evaluated by computer, using numerical integration methods. However, such an integration follows directly and therefore the probability of failure is calculated.

4-4 MONTE CARLO FAILURE ANALYSIS

The probability of failure of a system can be calculated by using the Monte Carlo analysis method discussed in Chapter 3. Failure must be defined, and then each Monte Carlo sample analysis must be evaluated to see if a failure has occurred. The probability of failure is obtained by dividing the total number of Monte Carlo sample analyses which experienced failure by the total number of Monte Carlo sample analyses.

Example 4-9

Consider the cantilever beam discussed in Examples 3-1 and 3-7. The moment induced load for 20 Monte Carlo samples is specified in Example 3-7. In Problem 3.15, 20 random values are generated for the yield stress of A-36 steel. Assume that the beam cross section has a section modulus which is deterministic and equal to 6 in.^{3}. Define failure to occur when the load-induced stress exceeds the material yield stress. Calculate the probability of failure.

The load-induced stress for the ith Monte Carlo sample is

$$^{(i)}f = \frac{^{(i)}M}{S}$$

For example, the load-induced stress for the first sample is

$$^{(1)}f = \frac{-(23.41)(12)}{(6)} = -46.82 \text{ ksi}$$

The following table gives the calculated load-induced stress for the 20 Monte Carlo examples.

Sample Number	Load Induced Stress (ksi)	Material Yield Stress (ksi)
1	−46.82	−48.18
2	−53.74	−51.36
3	−57.20	−52.95
4	−47.84	−48.66
5	−48.40	−48.91
6	−51.04	−50.12
7	−52.50	−50.78
8	−37.30	−43.82
9	−56.34	−52.54
10	−35.24	−42.88
11	−42.06	−46.00
12	−44.20	−46.99
13	−52.66	−50.86
14	−56.50	−52.62
15	−34.58	−42.58
16	−34.32	−42.45
17	−47.62	−48.55
18	−52.40	−50.74
19	−40.18	−45.14
20	−55.50	−52.16

Comparing the load-induced stress to the material yield stress for each sample it is possible to identify the total number of samples which indicated failure. Therefore, based on 20 samples the probability of failure is equal to $9/20 = 0.45$ or 45%.

Example 4-10

In Example 3-9, the collapse of the structure was assumed to occur when the maximum ground acceleration was greater than a_3. Now assume that a_3 is a uniform random variable with a mean of 125 cm/sec^2 and a coefficient of variation of 40%. Perform a Monte Carlo analysis to calculate the probability of structural collapse.

The limits of the uniform PDF for the collapse level acceleration are

$$\bar{a}_3 = \frac{a + b}{2} = 125 \text{ cm/sec}^2$$

$$\sigma_{a_3}^2 = \frac{(b - a)^2}{12} = 25.0 \text{ cm/sec}^2$$

$$\therefore \quad a = 168.30 \text{ cm/sec}^2$$

$$b = 81.7 \text{ cm/sec}^2$$

Using the first 20 numbers in column one of Table 3-1 the following table lists the corresponding 20 samples of collapse level acceleration. A sample-by-sample comparison of maximum ground acceleration and collapse level acceleration shows that no samples indicate failure. Therefore, based on a 20-sample Monte Carlo study the probability of collapse is 0/20 = 0.0.

Sample Number	Maximum Ground Acceleration (cm/sec²)	Collapse Level Acceleration (cm/sec²)
1	68.60	127.15
2	111.58	151.20
3	141.02	163.22
4	58.99	130.72
5	82.37	132.63
6	72.31	141.78
7	82.30	146.85
8	28.57	94.11
9	99.99	160.18
10	29.35	86.96
11	42.02	110.64
12	59.79	118.07
13	88.16	147.44
14	116.07	160.79
15	32.82	84.67
16	29.55	83.72
17	61.99	129.93
18	91.58	146.51
19	40.53	104.07
20	96.16	157.27

4-5 FACTORS OF SAFETY

In the previous sections of this chapter we defined the term probability of failure of a structure. In this section we discuss a term which is common in structural engineering, called the *factor of safety*. Confusion often exists because engineers try to relate these two terms and infer a value for one from the other. This is a mistake because the former term is consistent with use of probability in structural analysis whereas the latter term, even though it may be defined by using statistical quantities, is a deterministic quantity and is a term derived from classical deterministic structural design.

The *central factor of safety* has been previously defined and is

$$C_0 \equiv \frac{\bar{R}}{\bar{S}}$$

This factor of safety is obtained by dividing the mean resistance capacity (e.g., mean yield stress) by the mean induced loading response (e.g., mean load-induced member stress). It represents, in a general sense, the safety of the structure, but it fails to incorporate the uncertainty in the resistance and loading response. Recall that if a variable is deterministic and not random, then its standard deviation is zero. Therefore, if all variables describing the structure are deterministic, then the central factor of safety would indicate the safety of the structure.

A second commonly referred to safety factor is the *conventional safety factor* and is defined as

$$C_c = \frac{r_p}{s_q} \tag{4-30}$$

where r_p and s_q are defined as indicated in Figure 4-7. Consider first the resistance R part. Associated with any prescribed value of the resistance (e.g., the yield stress), one

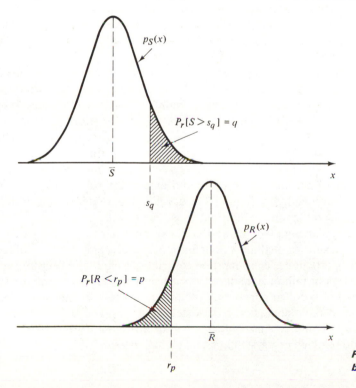

FIGURE 4-7. R and S probability density function.

can calculate the probability that the random variable, the resistance, will be less than the prescribed value. For example, the probability that the resistance (e.g., the yield stress) will be less than a prescribed value r_p is $\Pr[R \leq r_p] = p$ as indicated in Figure 4-7. Therefore, given a value of r_p and the PDF of R, one can obtain the value of p. For example, if the PDF is normal and r_p is equal to the mean $\bar{R}$ minus one standard deviation σ_R, then

$$r_p = \bar{R} - \sigma_R$$

and

$$p = 0.1587$$

Similarly, given a value of s_q, the load-induced response (e.g., load-induced stress), and the PDF of S, one can obtain the value of q, where q is the equal to $\Pr[S \geq s_q] = q$. (See Figure 4-7). For example, if the PDF is normal and s_q is equal to the mean $\bar{S}$ plus one standard deviation σ_s, then

$$s_q = \bar{S} + \sigma_S$$

and

$$q = 0.1587$$

Therefore, the conventional safety factor is obtained by dividing a number associated with resistance by a number associated with load-induced response, and these numbers can be seen to have a probabilistic interpretation. One can visualize the central factor of safety as a special case of the conventional factor of safety, where $r_p = \bar{R}$ and $s_q = \bar{S}$ and for normal PDF's $p = q = 0.50$.

These two common factor of safety terms, especially the latter, are often used to bridge the gap between simple deterministic building code or standards criteria and actual probabilistic structural analysis studies. Example 4-11 provides some insight into this development, which is beyond the scope of this introductory book.

Prior to concluding this chapter it is important to stress again that the structural engineer must define what is meant by failure. In the past, this fundamental concept has not been given the attention it deserves. Only after a clear and precise definition of failure has been made can one incorporate and extend the basic ideas presented here and calculate the probability of failure. The development of definitions of failure for different environmental loading conditions (e.g., earthquake or tornado loading) can be difficult, often involving planning and political input. However, the structural engineer is the key person for making the definition because only he has sufficient knowledge of the structure under consideration to make a meaningful decision.

After the definition of failure is made, then the acceptable level of probability of failure must be defined for design purposes. In the past, because of inadequate definitions of failure, considerable confusion has existed in attempts to define acceptable values for probability of failure. This confusion will decrease only when this difficulty is faced squarely and resolved in a systematic and logical manner.

Example 4-11

Consider the same problem as discussed in Example 4-1. Assume that

$$r_p = \bar{R} - \sigma_R = 47.9 - 3.3 = 44.6 \text{ ksi}$$

and

$$s_q = \bar{S} + \sigma_S = 36.0 + 7.2 = 43.2 \text{ ksi}$$

Calculate and discuss the conventional factor of safety.

Recall from Example 4-1 that

$$C_0 = \text{central safety factor} = \frac{47.9}{36.0} = 1.33$$

The conventional factor of safety for the given r_p and s_q is

$$C_c = \frac{r_p}{s_q} = \frac{44.6}{43.2} = 1.03$$

The value r_p can be visualized as a design yield stress value obtained by multiplying the mean yield stress by a stress reduction factor. That is,

$$r_p = \phi_R \bar{R}$$

where

$$\phi_R = \frac{r_p}{\bar{R}} = \frac{44.6}{47.9} = 0.93$$

Similarly, the value s_q can be visualized as a design load-induced stress value obtained by multiplying the mean load-induced stress by a load amplification factor. That is,

$$s_q = \phi_s \bar{S}$$

where

$$\phi_s = \frac{s_q}{\bar{S}} = \frac{43.2}{36.0} = 1.20$$

The present example therefore has a ϕ_R value which corresponds to a probability that the material yield stress being less than r_p is 15.87% and a ϕ_s value which corresponds to a probability that the load-induced stress being greater than r_q is 15.87%. The reader can see that it is possible to vary ϕ_R and ϕ_s such that different probabilities can be obtained. Note that in Example 4-1 we calculated the corresponding probability of failure for this problem as 6.7%.

4-6 REFERENCES AND ADDITIONAL READING

References

[4.1] BORGES, J. F., and M. CASTANHETA (1971): *Structural Safety*, Laboratorio National de Engenharia Civil, Lisbon, Portugal.

Additional Reading

BENJAMIN, J.R., and C.A. CORNELL (1970): *Probability, Statistics, and Decision for Civil Engineers*, McGraw-Hill Book Company, New York.

GHIOCEL, D., and D. LUNGU (1975): *Wind, Snow and Temperature Effects on Structures Based on Probability*, Abacus Press, Tunbridge Wells, Kent, England.

FREUDENTHAL, A.M., GARRELTS, J.M., and M. SHINOZUKA (1966): "The Analysis of Structural Safety," *Journal of the Structural Division*, ASCE.

ROBERTSON, L.E., and T. NAKA (1980): "Tall Building Criteria and Loading," American Society of Civil Engineers, New York, Chapter CL-2.

ELLINGWOOD, B. (1978): "Reliability Basis of Load and Resistance Factors for Reinforced Concrete Design," *NBS Building Science Series Report 110*, Center for Building Technology, National Bureau of Standards, Washington, D.C.

HAUGEN, E.B. (1968): *Probabilistic Approaches to Design*, John Wiley & Sons, Inc., New York.

RODIN, J. (1977): "Rationalisation of Safety and Serviceability Factors in Structural Codes," *CIRIA Report 63*, Construction Industry Research and Information Association, London.

PROBLEMS

4.1 The maximum fiber stress in a beam subjected to a bending moment is

$$f = \frac{Mc}{I}$$

where

M = applied moment

c = distance from neutral axis to extreme fiber

I = moment of inertia

Use your results from Problem 3.5 and assume that the maximum fiber stress has a normal PDF. Assume that failure occurs when the maximum fiber stress is equal to or greater than the material yield stress. The material yield stress (denoted as f_y) is considered to be a normally distributed random variable with mean and standard deviation equal to $\bar{f}_y$ and σ_{f_y}, respectively.

The load-induced stress and the yield stress are assumed to be independent random variables.

(a) Calculate the central factor of safety.

(b) Calculate $\bar{F}$, σ_F, and β if

$$F = f_y - f$$

(c) Explain how you would use β to calculate the probability of failure.

4.2 Consider Problem 3.6. The mean and variance of the load-induced support moment (denoted as M) was calculated in that problem. The moment capacity of the beam is denoted as M_c, and its mean and standard deviation are $\bar{M}_c$ and σ_{M_c}, respectively. The load-induced moment and the moment capacity of the beam are assumed to be independent normally distributed random variables.

Let

$\bar{M}$ = mean of load-induced moment = 50 kip-ft

σ_M = standard deviation of load-induced moment = 10 kip-ft

$\bar{M}_c$ = mean of beam's moment capacity = 60 kip-ft

σ_{M_c} = standard deviation of beam's moment capacity = 6 kip-ft

Failure is defined to occur when the load-induced moment exceeds the beam's moment capacity. Calculate the probability of failure.

4.3 In Problem 3.3 the support moment resulting from a concentrated and distributed load was considered and was equal to

$$M = \frac{Pl}{8} + \frac{\omega l^2}{12}$$

Let P and ω be independent normal random variables with the following statistical properties:

$\bar{P} = 1.0$ kip

$\sigma_p = 0.2$ kip

$\bar{\omega} = 0.10$ kip/ft

$\sigma_\omega = 0.02$ kip/ft

Consider l to be deterministic and equal to 12 ft. The beam which supports these loads has a cross section and material properties such that its calculated moment capacity (M_0) to resist the load-induced moment is

$\bar{M}_0$ = mean moment capacity = 3.00 kip-ft

σ_{M_0} = standard deviation of moment capacity = 0.30 kip-ft

Assume that M and M_0 are independent normally distributed random variables. Failure is defined to exist when

$$M \geq M_0$$

Calculate the probability of failure.

4.4   In Problem 4.3 the calculated moment capacity of the beam and its statistics were given. Now imagine that the moment capacity is defined as

$$M_0' = \phi M_0$$

where

$$\phi = \text{a capacity reduction factor}$$

The capacity reduction factor is a deterministic scalar which is used to reduce the calculated moment capacity of the beam for purposes of design.

(a)   Let $\phi = 0.90$. Calculate the mean and standard deviation of M_0' if the mean and standard deviation of M_0 have the values given in Problem 4.3.

(b)   Failure is now defined to occur when

$$M \geq M_0'$$

Use the statistics for M from Problem 4.3, and calculate the probability of failure. M_0' is assumed to have a normal PDF and is independent of M.

(c)   Repeat parts (a) and (b), but now let $\phi = 0.80$.

(d)   What observation can be made about the calculated probability of failure and the value of the capacity reduction factor?

4.5   Consider Problem 3.8 with

$$\bar{P} = 100 \text{ kips}$$
$$\sigma_p = 20 \text{ kips}$$

Assume that all members of the truss are steel and that their yield stress f_y is a normal random variable with the following statistics:

$$\bar{f}_y = \text{mean yield stress} = 47.9 \text{ ksi}$$
$$\sigma_{f_y} = \text{standard deviation of yield stress} = 3.8 \text{ ksi}$$

In Problem 3.8 the mean and standard deviation of the axial force in member CD of the truss (now denoted as F_{CD}) was calculated. Define

$$F_0 \equiv A f_y$$

where

$$F_0 = \text{axial force capacity of truss member } CD$$
$$A = \text{cross-sectional area of member } CD$$

Let f_y and F_{CD} be independent normally distributed random variables. The area A is assumed to be deterministic. Define failure to occur when the load-induced force is equal to or greater than the axial force capacity of the member; i.e.,

$$F_{CD} \geq F_0$$

(a) Calculate the minimum area A required for the probability of failure to be less than 0.10.

(b) Calculate the minimum area A required for the probability of failure to be less than 0.01.

4.6 Consider Problem 3.9 with

$$\bar{P} = 100 \text{ kips}$$
$$\sigma_p = 20 \text{ kips}$$

Assume that all members of the truss are steel and that their yield stress f_y is a normally distributed random variable with the following properties

$$\bar{f}_y = \text{mean yield stress} = 47.9 \text{ ksi}$$
$$\sigma_{f_y} = \text{standard deviation of yield stress} = 3.8 \text{ ksi}$$

Denote the axial force in truss member AF as F_{AF}. Define

$$F_0 = Af_y$$

where

$$A = \text{cross sectional area of member } AF$$
$$F_0 = \text{axial force capacity of truss member } AF$$

Let f_y and F_{AF} be independent normally distributed random variables. The area A is assumed to be deterministic. Define failure to occur when the load-induced force is equal to or greater than the axial force capacity of the member; i.e.,

$$F_{AF} \geq F_0$$

(a) Calculate the minimum area A required for the probability of failure to be less than 0.10.

(b) If the member AF were designed to have an area that was twice the value obtained in part (a), what would be the probability of failure of the design member?

4.7 Repeat Problem 4.2, but now assume that the load-induced moment and the moment capacity of the beam are independent log-normally distributed random variables.

4.8 Repeat Problem 4.5, but now assume that the load-induced force and the axial force capacity are independent log-normally distributed random variables.

4.9 Repeat Problem 4.6, but now assume that the load-induced force and the axial force capacity are independent log-normally distributed random variables.

4.10 Repeat Problem 4.2, but now assume that the load-induced moment and the moment capacity of the beam are independent uniformly distributed random variables.

4.11 Repeat Problem 4.5, but now assume that the load-induced force and the axial force capacity are independent uniformly distributed random variables.

4.12 Repeat Problem 4.6, but now assume that the load-induced force and the axial force capacity are independent uniformly distributed random variables.

5

Structural Loads

5-1 INTRODUCTION

One of the exciting aspects of being a structural engineer is the demand of facing the challenges required by society's need for new and different types of structures. Historically, this need has been met by the design and construction of such diverse structures as Roman bridges, Gothic cathedrals, sports coliseums, and skyscrapers. The analysis and design of any structure requires the identification of the important environmental factors which induce loads on the structure, and then the quantification of those loads. The continuous advancement of (1) our understanding of the physical factors which combine to produce such loads and (2) the analytical tools used to calculate structural response lead to structural designs which strive to satisfy the goals of society. The structural engineer's objective is a design which meets an acceptable level of safety while minimizing the use of material and financial resources.

The diverse kinds of structures which must be considered dictate corresponding diversities of loading. It is not our intent in this chapter to characterize all of the possible loads which can act on structures. Such a task would be immense and inappropriate for an introductory book on this subject. However, it is our intent to indicate how structural loading can be viewed in a more global sense. This is possible due to the statistical nature of the load occurrences and the historical manner in which engineers and scientists have acquired loading magnitude data. Therefore, while reading this chapter the reader is encouraged to keep this global view in mind even though the examples presented are only a small subset of all possible structural loads.

Viewing loads which act on a structure in a global sense it is possible to divide such loads into three basic types, as follows:

Loading Type I: These loads are described by using a statistical analysis of data obtained from measurements of load intensity and obtainable without regard for the time frequency of occurrence. The weight of the materials used in construction (dead loads) or the weight of the contents of a building (live loads) are examples of loads of this type.

Loading Type II: Loads which are described by using a statistical analysis of data obtained from measurements obtainable at prescribed periodic time intervals are of this type. Severe wind, snow, and storm-induced wave loads are examples of this kind of load.

Loading Type III: A description of loads of this type involves a statistical analysis of data obtained from infrequent measurements which are not obtainable at prescribed time intervals. The loading induced on a structure by earthquakes, tornadoes, and hurricanes are of this type.

5-2 LOADING TYPE I

Many loads which act on structures can be quantified by a direct application of the material presented in Chapter 2. The magnitude of these loads are random variables, and therefore one must establish their probability density (or distribution) function and numerical values for the associated PDF parameters. Values for the mean, standard deviation, and sometimes higher statistical moments must be estimated from load data. Type I load data is obtained without regard for the time frequency of occurrence. Stated in another way, we mean that one is concerned with the magnitude and sometimes the location of the load, but not how the magnitude and location vary with time.

An illustrative example of type I loads in the next section deals with dead and live loads.

5-3 LOADING TYPE I: DEAD AND LIVE LOADS

The two most common type I loads which structural engineers encounter are dead and live loads. These loads are defined as follows:

Dead load is the vertical load due to the weight of all permanent structural and nonstructural components of a structure, such as walls, floors, roof, beams, columns, and fixed service equipment.

Live load is the load superimposed by the use and occupancy of the structure, not including the time-varying loads discussed as type II and III loads later in this chapter, such as wind load, earthquake load, and snow load.

Dead loads are also referred to as *gravity loads*. The magnitude of this load is a random variable. This magnitude is reasonably predictable for new structures, where the geometry and materials of the structure are specified and equipment specifications are known. The variation in the load magnitude depends on the type of structure or nonstructural component, the material used in construction, and the degree and quality of field inspection utilized to ensure adherence to design specifications. Studies indicate that the unit weight of materials and the dead weight of structural components supplied by manufacturers provide good estimates of the mean value of dead loads. The coefficient of variation of the dead load has been studied, and for most design and construction in the United States it ranges from 6 to 13%. Therefore, a reasonable general estimate for the coefficient of variation of dead load is 10% in the absence of a detailed statistical study for a specific class of structures.

A linear statistical analysis which involves dead load requires the mean and standard deviation to be specified; therefore, the above discussion is sufficient. However, a Monte Carlo analysis requires quantification of the PDF of the dead load. The most common PDF selected for the dead load is the normal PDF. However, it is often desirable to utilize a log-normal PDF due to its ability to associate zero probabilities of occurrence with negative magnitudes of dead loads.

Live load is a function of the use and occupancy of the structure. Table 5-1 provides an illustration of this dependence and shows the minimum uniformly distributed and concentrated live loads recommended by the 1979 Uniform Building Code (UBC) [5.1].

The National Bureau of Standards (NBS) conducted an in-depth study of live loads in buildings in the United States [5.2]. This study surveyed private and government buildings to establish the magnitude and location of live loads. A review of a portion of the results of the study provides insight into the following statistical variations: (1) pounds per square foot live load magnitude, (2) pounds per square foot live load magnitude dependence on room usage, (3) pounds per square foot live load magnitude dependence on room area, (4) heaviest single live load magnitude, and (5) location of live loads within a room.

Figure 5-1 is a frequency histogram of the live load data obtained from a survey of 625 general and clerical offices in private office buildings. The histogram indicates a skew shape; therefore, a log-normal PDF and not a normal PDF can provide a good fit to the data. The coefficient of variation of the data is 5.5 psf divided by 9.0 psf, or 61%. The material in Chapter 2 and Figure 5-1 can be utilized to make probabilistic statements about the magnitude of the live load.

The magnitude of the live load is a function of the use of a room. Table 5-2 indicates the mean uniform live load and the corresponding coefficient of variation for several room usages. The statistical variation is large for all room types, and the difference between the mean live load magnitude for different room types can be very large (e.g., 8.7 psf for clerical and 24.9 psf for library).

Figures 5-2 through 5-4 and Table 5-3 indicate that the live load magnitude is also strongly dependent upon the size of the room. This data includes the previously

TABLE 5-1. Uniform Building Code Minimum Uniform and Concentrated Live Loads [5.1]

Use or Occupancy		Uniform Load (*psf*)	Concentrated Load (*lb*)
Category	*Description*		
1. Armories		150	0
2. Assembly areas and auditoriums and balconies therewith	Fixed seating areas	50	0
	Moveable seating and other areas	100	0
	Stage areas and enclosed platforms	125	0
3. Cornices, marquees, and residential balconies		60	0
4. Exit facilities, public		100	0
5. Garages	General storage and/or repair	100	
	Private pleasure car storage	50	
6. Hospitals	Wards and rooms	40	1000
7. Libraries	Reading rooms	60	1000
	Stack rooms	125	1500
Manufacturing	Light	75	2000
	Heavy	125	3000
8. Offices		50	2000
9. Printing plants	Press rooms	150	2500
	Composing and linotype rooms	100	2000
10. Residential		40	0
11. Rest rooms		≤ 50	
12. Reviewing stands, grand stands, and bleachers		100	0
13. Schools	Classrooms	40	1000
14. Sidewalks and driveways	Public access	250	
15. Storage	Light	125	
	Heavy	250	
16. Stores	Retail	75	2000
	Wholesale	100	3000

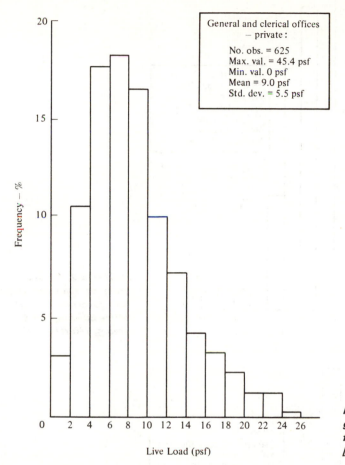

General and clerical offices
— private:

No. obs. = 625
Max. val. = 45.4 psf
Min. val. 0 psf
Mean = 9.0 psf
Std. dev. = 5.5 psf

FIGURE 5-1. Frequency histogram of live-load magnitude for private office buildings [5.2].

TABLE 5-2. Influence of Room Use on Live Loads in Private Office Buildings [5.2]

Room Use	Mean Live Load (psf)	Coefficient of Variation of Live Load
General	8.7	0.56
Clerical	10.0	0.69
Lobby	4.6	1.04
Conference	6.1	0.82
File	24.4	0.79
Storage	15.5	0.89
Library	24.9	0.39
All Rooms	9.7	0.84

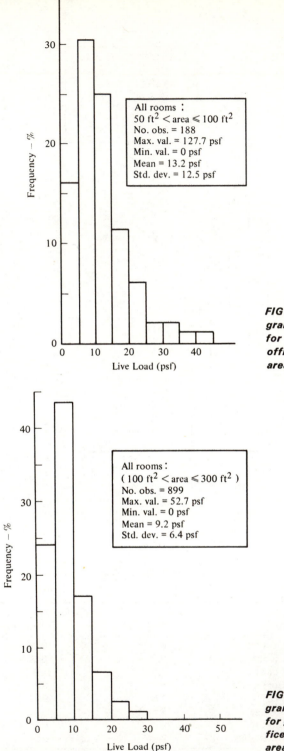

All rooms :
$50 \text{ ft}^2 < \text{area} \leqslant 100 \text{ ft}^2$
No. obs. = 188
Max. val. = 127.7 psf
Min. val. = 0 psf
Mean = 13.2 psf
Std. dev. = 12.5 psf

Frequency – %

Live Load (psf)

FIGURE 5-2. Frequency histogram of live-load magnitude for government and private office buildings (50 ft² < room area ≤ 100 ft²) [5.2].

All rooms :
$(100 \text{ ft}^2 < \text{area} \leqslant 300 \text{ ft}^2)$
No. obs. = 899
Max. val. = 52.7 psf
Min. val. = 0 psf
Mean = 9.2 psf
Std. dev. = 6.4 psf

Frequency – %

Live Load (psf)

FIGURE 5-3. Frequency histogram of live-load magnitude for government and private office buildings (100 ft² < room area ≤ 300 ft²) [5.2].

140

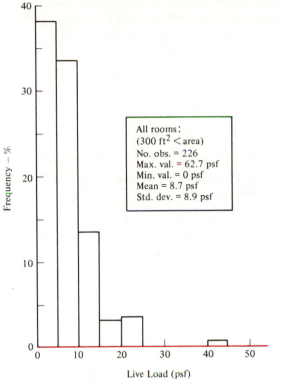

FIGURE 5-4. Frequency histogram of live-load magnitude
for government and private
office buildings (room area >
300 ft²) [5.2].

TABLE 5-3. Influence of Room Area on Live Loads in Private
and Government Office Buildings [5.2]

Room Area (ft^2)	Number of Rooms	Mean Live Load (psf)	Coefficient of Variation of Live Load
0 to 50	41	17.4	1.00
50+ to 100	188	13.2	0.95
100+ to 300	899	9.2	0.70
Greater than 300	226	8.7	1.02

described private office building data plus additional government office building data
for 520 rooms. The government office building data resulted in a slightly greater mean
live load magnitude and coefficient of variation than the private office building data
(10.3 psf vs. 9.7 psf and 0.90 vs. 0.84). This reduction in live load magnitude with floor
area is recognized in building codes. The Uniform Building Code [5.1] allows the
values given in Table 5-1 to be reduced when the area of the floor supported by the
horizontal or vertical structure member exceeds 150 sq ft.

The percent reduction R is

$$R = 0.08A \qquad (5\text{-}1)$$

The reduction shall not exceed 40% for horizontal members (e.g., beams) or vertical members (e.g., columns) receiving load from one level only, or 60% for other vertical members, or R as determined by the following formula:

$$R = 23.1 \left(1 + \frac{D}{L}\right) \qquad (5\text{-}2)$$

where

R = reduction (percent)

A = area of floor supported by the structural member

D = dead load per square foot of area supported by the member

L = live load per square foot of area supported by the member (see Table 5-1)

The live load data previously discussed is uniformly distributed live load data specified in units of pounds per square foot. This data represents an averaged load obtained by dividing the total live load in a room by the total area of that room. It is

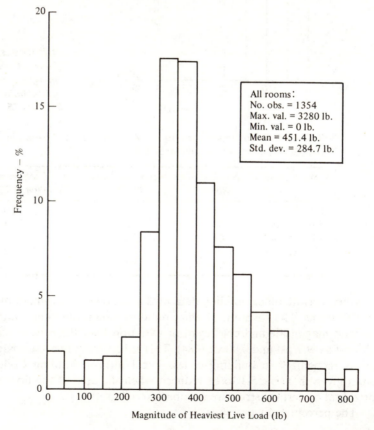

All rooms:
No. obs. = 1354
Max. val. = 3280 lb.
Min. val. = 0 lb.
Mean = 451.4 lb.
Std. dev. = 284.7 lb.

Magnitude of Heaviest Live Load (lb)

FIGURE 5-5

informative to consider other data that were obtained in the survey. For each room surveyed the single heaviest object was weighted; Figure 5-5 shows a histogram for the magnitude of the heaviest live load obtained, using all 834 private office building rooms and 520 government office building rooms. This histogram indicates that the mean heaviest live load was 451.4 lb, and the coefficient of variation was 63%. Table 5-1 indicates the magnitude of concentrated live loads considered by the Uniform Building Code. (Note: The concentrated loads given by the UBC are to be distributed over a $2\frac{1}{2}$ sq ft area.)

The spatial location of live loads within any single room does not tend to be uniform over the room area. Figure 5-6 and Table 5-4 indicate that most live loads within a room tend to be near the walls. For example, 75.6% of the live loads are within 2 ft of the walls. A visual inspection of the room in which the reader is seated will probably confirm this common tendency to locate book shelves, filing cabinets, etc., near walls.

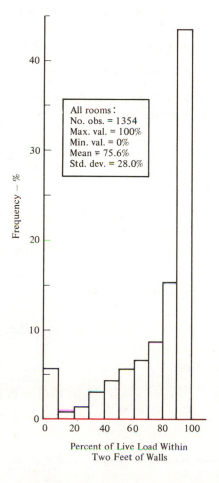

All rooms:
No. obs. = 1354
Max. val. = 100%
Min. val. = 0%
Mean = 75.6%
Std. dev. = 28.0%

Percent of Live Load Within
Two Feet of Walls

FIGURE 5-6

TABLE 5-4. Spatial Distribution of Live Load Objects Within Rooms [5.2]

Object	Mean Percent of Objects Within 2 ft of Walls (Government Offices)	Mean Percent of Objects Within 2 ft of Walls (Private Offices)
Desk	55.1	55.5
Table	69.8	61.8
Cabinet	95.7	84.6
Shelving	72.5	81.7
Seating	35.0	31.8
Miscellaneous	65.2	54.7

Example 5-1

Consider the live load data listed in Table 5-2 for different room usage. Assume that the PDF of the live load magnitude is log-normal. Calculate the probability that live load in a general office will exceed 10 psf.

The mean and coefficient of variation of the live load are 8.7 psf and 0.56, respectively. The PDF of the live load (denoted as L) is, from Eq. (2-84),

$$p(l) = \frac{1}{l\sigma_Y\sqrt{2\pi}} \exp\left[-\frac{1}{2}\left(\frac{\ln l - \bar{Y}}{\sigma_Y}\right)^2\right]$$

where

$$\sigma_Y^2 = \ln(\rho_l^2 + 1) = \ln[(0.56)^2 + 1] = 0.272$$

$$\bar{Y} = \ln \bar{l} - \frac{1}{2}(\sigma_Y^2) = \ln 8.7 - \frac{0.272}{2} = 2.027$$

From Eq. (2-87) it follows that

$$U^* = \left[\frac{\ln 10.0 - 2.027}{0.522}\right] = 0.528$$

Therefore,

$$\Pr[L > 10.0] = \Pr[U^* > 0.528]$$
$$= 1 - \Pr[U^* \le 0.528]$$
$$= 1 - 0.70 = 0.30, \text{ or } 30\%$$

Example 5-2

Consider the live load data listed in Table 5-3 for different room areas. Assume that the PDF of the live load magnitude is log-normal. Calculate the probability that the live load will exceed 15 psf for room areas 50+ to 100 and 100+ to 300.

Consider first the room area range 50+ to 100 sq ft, with a mean and coefficient of variation equal to 13.2 psf and 0.95, respectively. It follows from Eqs. (2-85) and (2-86) that

$$\sigma_Y^2 = \ln[(0.95)^2 + 1]$$

and

$$\bar{Y} = \ln 13.2 - \frac{0.643}{2} = 2.259$$

Finally, from Eq. (2-87),

$$U^* = \left(\frac{\ln 15.0 - 2.259}{0.802}\right) = 0.560$$

and

$$\Pr[L > 15.0 \text{ psf}] = 1 - \Pr[U^* \le 0.560] = 0.29$$

Similarly, for the room area range 100+ to 300 sq ft it follows that

$$\sigma_y^2 = \ln[(0.70)^2 + 1] = 0.399$$

$$\bar{Y} = \ln 9.2 - \frac{0.399}{2} = 2.020$$

$$U^* = \left(\frac{\ln 15.0 - 2.020}{0.632}\right) = 1.089$$

and

$$\Pr[L > 15.0 \text{ psf}] = 1 - \Pr[U^* \le 1.089] = 0.14$$

5-4 LOADING TYPE II

Many kinds of environmental loads have a common feature, namely, that they can be described by using data which can be obtained at prescribed periodic time intervals. For example, statistical data can be obtained for any year for the maximum depth of snow for the month of January or the maximum ocean wave height during the month of September. Year in and year out it is possible to obtain such data. Therefore, once a prescribed periodic time interval is established and the type of data to be collected identified, one is able to systematically gather data for statistical analysis. Structural loads that correspond to environmental conditions offering such regularity are referred to as *loading type II loads*.

As an illustrative example of type II loads we discuss wind loading in the next two sections.

5-5 LOADING TYPE II: DETERMINATION OF WIND SPEEDS

The wind is an environmental element which induces loads on structures. These loads usually require the structual engineer to address the following design conditions:

1. System and structural component strength (e.g., shear, axial, and bending capacity of beams and columns compared with load-induced shear, axial, and bending loads).
2. Nonstructural component strength and support (e.g., required window thickness and attachments).
3. Human comfort (e.g., building floor motions above human perception level or wind flow causing inconvenience through plaza areas).
4. Instability (e.g., the possible tuning of the vibrational characteristics of the wind and transmission lines or pollutant discharge stacks).

Such wind design considerations constitute a continuously evolving area of engineering study and research. This general area is known as *wind engineering* and a comprehensive presentation of the subject is beyond the scope of this introductory book. The reader is referred to a tall building design monograph chapter on wind loading [5.3] and to a recent book by Simiu and Scanlan [5.4] as good starting points for an in-depth study of wind engineering.

Our objective in this section is to provide some basic definitions of important wind engineering terms, to discuss how wind-speed data is obtained and analyzed, and to illustrate how structural engineers can describe the lateral and vertical spatial variations of wind speeds. Our objective in the next section is to carry this process one step further and illustrate how one can calculate the wind pressures on structural components and the total lateral wind force acting on a structure. Therefore, these two sections develop an introductory scenario describing the basic parts of the wind design process.

Thom [5.5] assimilated maximum annual wind-speed data at 141 locations in the United States and completed a statistical analysis of the data. Wind speed was measured by using an instrument called an *anemometer*. An anemometer usually has a propeller that rotates in the wind, and the amount of air mass passing the anemometer can be calibrated for one complete revolution of an anemometer blade. Therefore, one revolution of the blade corresponds to a distance of wind particle movement. After the blade has completed a certain number of revolutions it is possible to establish that the wind particle travel distance would be equal to 1 mile. The basic weather station anemometer used by the National Weather Service records a mark on a paper recorder every time the anemometer has completed the number of revolutions corresponding to 1 mile of wind mass passage. The paper or tape in the recorder moves at a constant speed, and therefore the elapsed time between marks is known. When this elapsed time is divided into the distance of 1 mile, one obtains an average wind speed in miles per hour for the wind which passed the anemometer site during the time between successive marks on the drum. After studying all anemometer records for a calendar year, it is possible to determine the shortest elapsed time between marks and therefore the annual fastest mile speed. This wind speed is called the *annual fastest-mile wind speed*. For each of the 141 U.S. stations, Thom collected this periodic data and for each station he statistically analyzed the data to determine a probabilistic model of wind speed.

Consider one station and the data that existed for that station. One data point existed for each year of record. Using this data the probability density function which Thom determined to be most appropriate was the Fréchet. The equation [see Eq. (2-65)] for this probability density function is

$$p(v) = \left(\frac{a}{b}\right)\left(\frac{b}{v}\right)^{a+1} \exp\left[-\left(\frac{b}{v}\right)^{a}\right] \qquad \begin{array}{l} v \geq 0 \\ a, b > 0 \end{array} \tag{5-3}$$

The values for the parameters a and b are determined by using the annual fastest-mile wind-speed data at the station under study. The probability of having an annual fastest-mile wind speed equal to or less than a set value of wind speed can be directly obtained from the Fréchet probability distribution function:

$$P(v) = \exp\left[-\left(\frac{b}{v}\right)^{a}\right] \tag{5-4}$$

where the magnitude of the wind speed is denoted as v. The probability that the wind speed for any single year will exceed the set value of wind speed is obtained by using Eq. (5-4) and is

$$\Pr[V > v] = 1 - P(v) \equiv t \tag{5-5}$$

where the symbol t is defined for notational convenience. Now consider the random variable, denoted as $T(v)$, which is defined as the number of years between those years where the wind speed is greater than v. This random variable is referred to as the *wind-speed return period*. The probability that the return period will be 1 year is

$$\Pr[T(v) = 1] = t \tag{5-6}$$

and the probability that it will be 2 years is

$$\Pr[T(v) = 2] = (1 - t)t \tag{5-7}$$

where the $(1 - t)$ term is the probability that the wind speed will not exceed v in the first year, and the term t is the probability that it will exceed v in the second year. It therefore follows that the probability that the return period will be exactly equal to i years is

$$\Pr[T(v) = i] = (1 - t)^{i-1}t \qquad (5\text{-}8)$$

The mean return period can now be calculated and is

$$E\langle T(v)\rangle = \bar{T} = \text{mean return period}$$
$$= (1) \Pr[T(v) = 1] + (2) \Pr[T(v) = 2] + \ldots$$
$$= \sum_{i=1}^{\infty} i \Pr[T(v) = i]$$
$$= \sum_{i=1}^{\infty} i(1 - t)^{i-1}t \cong \frac{1}{t} \qquad (5\text{-}9)$$

This series can be shown to be approximately equal to $1/t$ with less than a 4% error when the mean return period is 50 years or greater. Using Eqs. (5-5) and (5-9) it is clear that the mean return period and the annual fastest-mile wind-speed distribution function are related by the formula

$$\bar{T} = \frac{1}{1 - P(v)} \qquad (5\text{-}10)$$

or, alternatively,

$$P(v) = 1 - \left(\frac{1}{\bar{T}}\right) \qquad (5\text{-}11)$$

Using the data for each of the 141 stations, Thom determined the numerical values of the parameters in the wind-speed probability distribution function, Eq. (5-4). For a specified mean return period Thom then used Eq. (5-11) to calculate the value of the probability distribution function and then the value of wind speed corresponding to this distribution function value. Thom performed this procedure for each of the 141 stations and then constructed a map of the United States that indicated the geographic variation of wind speed for the specified mean return period. Figure 5-7 shows such a map for a 50 year return period. The term *mean recurrence interval* is often used instead of mean return period, but the meaning of both terms is the same. Such a map is referred to as a *wind-speed zoning map*.

The wind zoning concept has been expanded upon by restricting the size of a geographic region to a smaller area. Such a reduced area zoning is called *microzonation* [5.6], and Figure 5-8 shows a wind-speed map for a microzoning study of the greater Los Angeles area. It is apparent from a comparison of Figures 5-7 and 5-8 that microzoning incorporates the local meteorological conditions of the area in a more precise manner, and that these conditions can in some geographic areas have a significant influence on the estimated wind speed for a given return period.

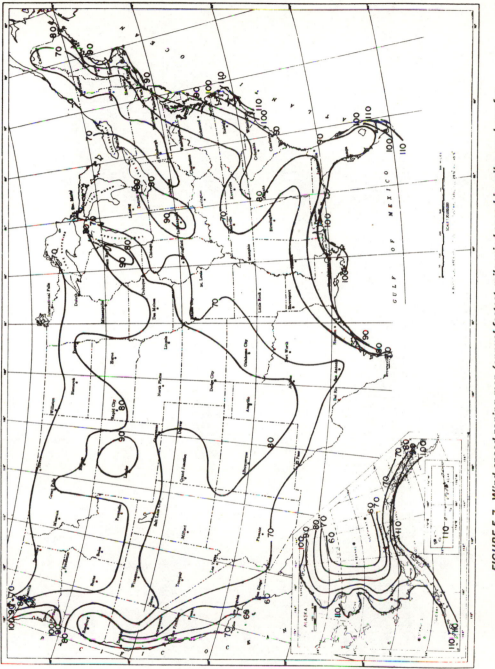

FIGURE 5-7. Wind-speed zoning map (annual fastest-mile wind speed in miles per hour for an open-country exposure and at a height of 30 ft: 50-year return period) [5.2].

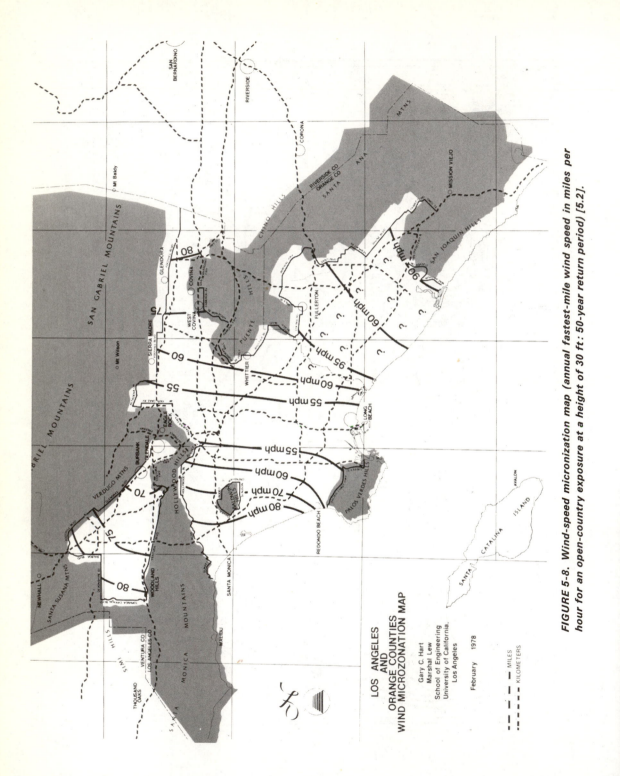

FIGURE 5-8. *Wind-speed micronization map (annual fastest-mile wind speed in miles per hour for an open-country exposure at a height of 30 ft: 50-year return period) [5.2].*

The study by Thom developed a data set for each station that was composed of the largest fastest-mile wind-speed value for each year. Usually, for any site the wind speed varies significantly from month to month. Therefore, it is possible to periodically select the largest fastest-mile wind-speed value for each month instead of for each year. If this is done, then instead of one data set there will exist 12 data sets (i.e., one data set for each calendar month). Thom analyzed the data set for the fastest-mile wind speed on an annual basis and established values for *a* and *b* in Eq. (5-3). It is possible to alternatively analyze the data set for any selected month and therefore establish values for *a* and *b* in Eq. (5-3). If this is done, then the engineer can calculate the wind speed for any selected month corresponding to a specified mean return period. Table 5-5 shows the results of such a study, and it is apparent that the wind speed for a given return period can vary considerably from month to month.

The wind usually has a directional dependence resulting from basic meterological or geographical considerations. Therefore, the structural engineer may desire a study of the directional dependence of wind speeds for such special problems as wind flow through plaza areas or wind loads on long, narrow buildings. The basic approach used by Thom can also be followed in such a study. However, instead of one largest, fastest-mile wind speed being selected for each year and being added to a data set, one largest, fastest-mile wind speed is selected for each year for each of the eight commonly used compass directions; thus, eight data sets are created for analysis. Once a data set is created, the same Thom data analysis procedure is followed, and for a given compass direction the engineer can calculate the wind speed for a specified mean return period. Table 5-6 shows the results of such a study, and it is apparent that for the site selected the directional variation is significant.

TABLE 5-5. Monthly Variation of Wind Speed (Los Angeles International Airport) [5.6]

Month	Return Period in Years				
	2	10	25	50	100
All Months	52.5*	68.2	76.1	82.0	87.8
Jan	41.0	60.5	70.3	77.5	84.8
Feb	41.8	61.4	71.3	78.6	85.9
Mar	44.1	61.0	69.5	75.8	82.1
Apr	40.5	53.5	60.1	64.9	69.8
May	33.2	48.0	57.1	67.7	73.0
June	26.8	35.0	39.1	42.1	45.1
July	24.2	31.0	34.5	37.1	39.6
Aug	25.1	32.5	36.3	39.0	41.8
Sep	25.5	33.8	38.0	41.0	44.1
Oct	34.7	53.5	62.9	69.9	76.9
Nov	38.2	52.6	59.9	65.3	70.7
Dec	40.5	60.8	71.0	78.5	86.0

*Wind speeds are specified by fastest mile, in miles per hour, open-country exposure, and at 30 ft elevation.

TABLE 5-6. Directional Variation of Wind Speed (Los Angeles International Airport) [5.6]

Direction	Wind Speed (50 year return period)
North	86*
Northeast	68
East	52
Southeast	56
South	68
Southwest	58
West	63
Northwest	71

*Wind speeds are specified by fastest mile, in miles per hour, open-country exposure, and at 30 ft elevation.

Wind speed varies with height. This variation manifests itself in two forms. First, mean wind speed increases with height because of the friction forces induced by the roughness of the earth's surface. The closer the air is to the earth's surface, the more this roughness decreases the mean speed of the air flowing over the earth. Second, the variations in wind speed about the mean wind speed are larger near the surface because of this same surface roughness. At a height called the *gradient height* (H_G), the frictional influence of the surface roughness is assumed to be negligible. The wind speed at this height is called the *gradient wind speed* (V_G). The wind speed at heights greater than the gradient height is assumed to be equal to the gradient wind speed. The gradient wind speed is influenced by regional meterological conditions.

The variation of *mean wind speed* with height is described by using either a power law or logarithmic law profile [5.4]. The former has, historically, more commonly been used, whereas the latter has certain theoretical advantages. The power law profile describes the mean wind speed vs. height, using the formula

V_H = mean wind speed at height H above the ground

$$= \left(\frac{H}{H_G}\right)^\alpha V_G \tag{5-12}$$

As an illustration of this variation, Figure 5-9 shows the variation for three classifications of ground roughness [5.7]. Table 5-7 lists values for the gradient height and the power law coefficient α. The numerical values for the gradient height and the power law coefficient for a given ground roughness are not universal. However, the concept of roughness classification is important, and ambiguity in the definition of the roughness for each class should not deter the reader from the fundamental significance of the benefit and need to classify terrain and relate the classifications to H_G and α values. Reflecting on Eq. (5-12) it is apparent that, given the classification of a site and the H_G and α values (e.g., Table 5-7), one can determine the mean wind speed at a given height once the gradient wind speed is established. Therefore, attention is now directed to the determination of the gradient wind speed at the site.

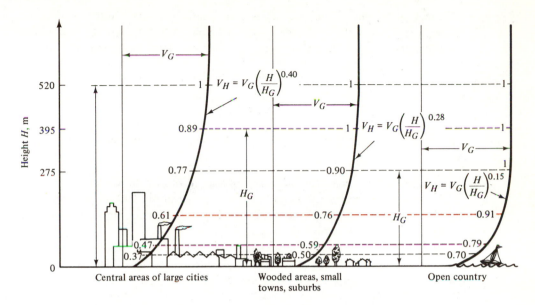

FIGURE 5-9. *Mean wind-speed variation with height [5.7].*

TABLE 5-7. Mean Wind-Speed Power Law Coefficients [5.7]

Surface Roughness Classification or Site Exposure	Gradient Height (H_G)	Power Law Coefficient (α)
Central area of large cities	520 m (1,500 ft)*	0.40 (1/3)*
Wooded areas, small towns, or suburbs	395 m (1,200 ft)	0.28 (2/9)
Open country	275 m (900 ft)	0.16 (1/7)

*Numbers in parentheses ( ) are current ANSI A58.1, 1972 recommended values [5.9].

The wind-speed zoning map established for a reference height (e.g., 10 m) the annual fastest-mile wind speed. As indicated in the definition of the term *fastest-mile wind speed*, this wind speed is an average wind speed over the time interval necessary for 1 mile of wind to pass the site. For example, if the fastest-mile wind speed is 60 mph, the averaging time is 1 min. Therefore, using Eq. (5-12) and a wind-speed zoning map, it follows that

V_{10} = fastest-mile wind speed at a 10 m height, obtained from a wind-speed zoning map

$$= \left(\frac{10}{H_G}\right)^\alpha V_G \qquad (5\text{-}13)$$

Thom's fastest-mile wind speeds are for a Table 5-7 open-country surface roughness classification, and therefore the gradient height is 275 m and the power law coefficient is 0.16. The gradient wind speed then follows from Eq. (5-13) and is equal to

$$V_G = \text{gradient-height wind speed}$$

$$= \frac{V_{10}}{(\frac{10}{275})^{0.16}} \qquad\qquad (5\text{-}14)$$

The mean wind speed at a desired structural site is determined by using the following steps:

Step 1. Locate the site on a wind-speed zoning map, and determine the wind speed at a height of 10 m for the prescribed mean return period.

Step 2. Calculate the gradient wind speed for an open-country exposure at the site, using Eq. (5-14).

Step 3. Establish the actual surface roughness classification for the structural site, and use Table 5-7 to define the actual gradient wind-speed height and power law coefficient.

Step 4. Calculate the mean wind speed at any height H for the prescribed mean return period, using Eq. (5-12) with the H_G and α from step 3 and the V_G from step 2.

Example 5-3

Ten years of annual fastest-mile wind speed have been obtained at a specific site. The data is as follows:

Year	Fastest Mile Wind Speed (mph)
1971	80.5
1972	75.7
1973	69.5
1974	82.0
1975	75.0
1976	72.0
1977	65.0
1978	84.0
1979	81.0
1980	67.0

Assume that the annual fastest-mile wind-speed data has a Fréchet PDF.

(a) Calculate the sample mean and standard deviation of the annual fastest-mile wind speed.

(b) Assume that this wind speed data has a Fréchet PDF. Calculate the probability distribution function of the wind speed.

(c) Calculate the probability that the annual fastest-mile wind speed will be greater than 80 mph in *any single year*.

(d) What is the mean return period for annual fastest-mile wind speeds greater than 80 mph?

Part (a) solution:

$$^s\bar{V} = \frac{1}{10}\sum_{i=1}^{10} V_i = \frac{1}{10}[80.5 + 75.7 + \ldots + 67.0] = 75.17 \text{ mph}$$

$$^s\sigma_v^2 = \frac{1}{10}\sum_{i=1}^{10}(V_i - {^s\bar{V}})^2 = \frac{1}{10}[(80.50 - 75.17)^2 + (75.70 - 75.17)^2$$

$$+ \ldots + (67.0 - 75.17)^2]$$

$$= 39.94 \text{ (mph)}^2$$

$$\therefore \quad {^s\sigma_v} = 6.32 \text{ mph}$$

Part (b) solution:

The coefficient of variation p_v is equal to

$$p_v = \frac{6.32}{75.17} = 0.084$$

It follows from Eqs. (2-66) and (2-67) that

$$p_v = \frac{\sigma_v}{\bar{V}} = \frac{\{\Gamma[1 - (2/a)] - \Gamma^2[1 - (1/a)]\}^{1/2}}{\Gamma[1 - (1/a)]}$$

Squaring both sides and rearranging terms, it follows that

$$1 + p_v^2 = \frac{\Gamma[1 - (2/a)]}{\Gamma^2[1 - (1/a)]}$$

Using $p_v = 0.084$, it follows from trial and error that

$$a \cong 16$$

Finally, from Eq. (2-66) it follows that

$$\bar{V} = 75.17 = b\Gamma[1 - (\tfrac{1}{16})]$$

and

$$b = 72.27$$

Therefore,

$$P(V \leq v) = \exp\left[-\left(\frac{72.27}{v}\right)^{16}\right]$$

Part (c) solution:

$$\Pr[V \le 80 \text{ mph}] = \exp\left[-\left(\frac{72.27}{80.00}\right)^{16}\right] = 0.821$$

Therefore, the probability that $V > 80$ mph for any single year is

$$\Pr[V > 80 \text{ mph}] = 1 - \Pr[V \le 80 \text{ mph}]$$
$$= 1.000 - 0.821 = 0.179$$

Part (d) solution:

$$\bar{T} = \text{mean return period} = \frac{1.0}{\Pr[V > 80 \text{ mph}]}$$

$$= \left(\frac{1}{0.179}\right) = 5.6 \text{ years}$$

Example 5-4

Assume that the annual fastest-mile wind-speed data given in Example 5-3 was obtained at a height of 10 m and for an open-country exposure. Also assume that a 50 year mean return period is to be used for design purposes.
  (a) Calculate the annual fastest-mile wind speed for the 50 year return period.
  (b) What is the corresponding annual fastest-mile wind speed at the gradient height (i.e., V_G at H_G)?

Part (a) solution:

A 50 year return period corresponds to a probability of occurrence for any single year of 0.02, or stated alternatively,

$$P[V > v] = 0.02 = \frac{1}{\bar{T}}$$

It follows that

$$P[V \le v] = 1 - 0.02 = 0.98$$

and

$$P[V \le v] = 0.98 = \exp\left[-\left(\frac{72.27}{v}\right)^{16}\right]$$

Therefore,

$$v = 92.23 \text{ mph}$$

Part (b) solution:

The open-country gradient height and power law coefficient from Table 5-7 are

$$H_G = 275 \text{ m}$$

$$\alpha = 0.16$$

The gradient wind speed V_G is, from Eq. (5-12),

$$V_G = \frac{V_H}{(H/H_G)^\alpha} = \frac{92.23}{(\frac{10}{275})^{0.16}} = 156.7 \text{ mph}$$

Example 5-5

The annual fastest-mile wind speed at the gradient height from Example 5-4 is assumed to be the same as the annual fastest-mile gradient wind speed at a building site located nearby, but for a suburban site exposure.

(a) Calculate the annual fastest-mile wind-speed profile at the building site over the height of the building. Assume that the building height is 30 m.

(b) If the building site exposure corresponded to that of an urban area or a large city, then define the wind-speed profile.

Part (a) solution:

For a surburban building site exposure,

$$H_G = 395 \text{ m}$$

$$\alpha = 0.28$$

Therefore, using from Example 5-4 that

$$V_G = 156.7 \text{ mph}$$

it follows that

$$V_H = 156.7 \left(\frac{H}{395}\right)^{0.28}$$

Part (b) solution:

For an urban building site exposure,

$$H_G = 520 \text{ m}$$

$$\alpha = 0.40$$

Therefore,

$$V_H = 156.7 \left(\frac{H}{520} \right)^{0.40}$$

Example 5-6

Table 5-5 presents the monthly variation in wind speed for a given return period. Assume that the fastest-mile wind-speed data for the month of December has a Fréchet PDF.

(a) Use the results shown in Table 5-5 to define the December fastest-mile wind-speed probability distribution function.

(b) Calculate the probability that the fastest-mile wind speed in any single month of December will exceed 50 mph. What is the return period of the fastest-mile wind speed exceeding 50 mph in December?

Part (a) solution:

The Fréchet probability distribution function [see Eq. (5-4)] is

$$\Pr[V \leq v] = \exp\left[-\left(\frac{b}{v}\right)^a\right]$$

Therefore,

$$\Pr[V > v] = 1 - \exp\left[-\left(\frac{b}{v}\right)^a\right] = \frac{1}{T}$$

From Table 5-5 it follows that, for December data,

$$\Pr[V > 40.5 \text{ mph}] = \frac{1}{T} = \frac{1}{2 \text{ years}}$$

and

$$\Pr[V > 60.8 \text{ mph}] = \frac{1}{T} = \frac{1}{10 \text{ years}}$$

Thus, two equations for the parameters a and b exist and are

$$\Pr[V > 40.5 \text{ mph}] = 1 - \exp\left[-\left(\frac{b}{40.5}\right)^a\right] = 0.50$$

and

$$\Pr[V > 60.8 \text{ mph}] = 1 - \exp\left[-\left(\frac{b}{60.8}\right)^a\right] = 0.10$$

Finally, solving the above, one obtains

$$a = 4.52$$

$$b = 37.31 \text{ mph}$$

The probability distribution function for December fastest-mile wind speeds is

$$P(v) = \exp\left[-\left(\frac{37.31}{v}\right)^{4.52}\right]$$

Part (b) solution:

Using the above and $v = 50$ mph, one obtains

$$P(v = 50 \text{ mph}) = \exp\left[-\left(\frac{37.31}{50.00}\right)^{4.52}\right] \doteq 0.77$$

The probability that any single December will have a wind speed exceeding 50 mph is 23%; therefore, the mean return period is 4.3 years (i.e., 1/0.23).

5-6 LOADING TYPE II: DETERMINATION OF WIND LOADS

In the previous section we discussed the determination of wind speeds at the site of a structure. Wind speeds are related to the wind pressure acting on structures and therefore represent an important component of wind-force determination. The other components include the geometry of the structure, the vibrational characteristics of the structure, and the mass density of the air. Prior to discussing these components, it is appropriate to provide an indication of the general magnitude of wind pressures that act on structures.

No single code or standard is used for wind design in the United States. Several candidate codes and standards exist, and local jurisdictions select the one they believe to be best for the structural engineering of their community. However, it is possible to obtain a general indication of the magnitude of wind pressures by studying a single candidate code. For example, consider the 1979 Uniform Building Code [5.1] wind pressures. Figure 5-10 shows a map of the continental United States, which identifies seven wind zones. These wind zones reflect the general geographical variation of wind speeds in much the same way as the wind-speed zoning map. The indicated wind pressures correspond to pressures at a height of 30 to 49 ft. The wind pressures are less than noted for heights below 30 ft and greater for heights above 50 ft. This variation is as indicated in Table 5-8.

Consider now the basic component parts which are combined so as to determine the wind loads on structures. For illustrative purposes imagine that a specific point

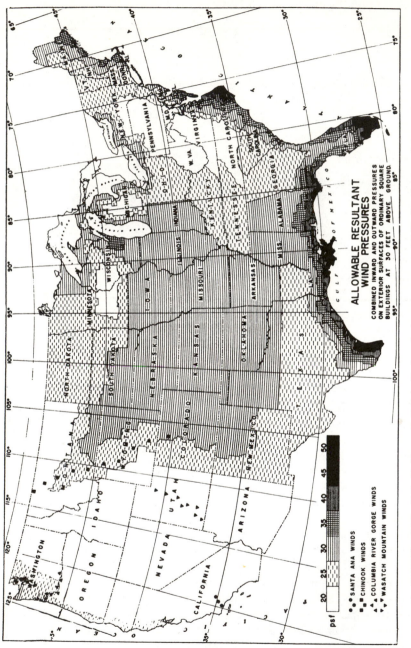

FIGURE 5-10. U.S. Uniform Building Code wind-pressure map [5.1]. (Reproduced from the 1979 edition of the Uniform Building Code, copyright 1979, with permission of the publisher, the International Conference of Building Officials.)

TABLE 5-8. Uniform Building Code Pressure Variation with Height [5.1]

Height Zones (ft)	Wind-Pressure Map Areas (psf)						
	20	25	30	35	40	45	50
Less than 30	15	20	25	25	30	35	40
30 to 49	20	25	30	35	40	45	50
50 to 99	25	30	40	45	50	55	60
100 to 499	30	40	45	55	60	70	75
500 to 1199	35	45	55	60	70	80	90
1200 and over	40	50	60	70	80	90	100

has been identified on the exterior face of a building and that one desires to know the wind pressure at that point. The wind pressure per unit area can be represented symbolically as the result obtained by multiplying the following three component parts:

$$\text{wind pressure} = (\text{gust factor})(\text{pressure coefficient})(\text{dynamic wind pressure})$$

The *dynamic wind pressure* incorporates the wind-speed data obtained from a zoning map and the mass density of the air; the *pressure coefficient* incorporates the geometry, or shape, of the structure and the fluid mechanics of how the wind flow strikes and flows around the structure; the *gust factor* incorporates the vibrational characteristics of the structure (e.g., its natural period of vibration and damping) and the turbulence characteristics of the wind. A detailed discussion of each of these components is contained in Simiu and Scanlan [5.4]. However, it is possible to provide an introduction here, and the remainder of this section is intended to provide such an insight.

The dynamic wind pressure is directly obtained from the mean wind speed at the site of the structure, for the height under consideration, using the equation

$$q_D(H) = \text{dynamic wind pressure at height } H$$

$$= \frac{\rho V_H^2}{2} \tag{5-15}$$

where

$$\rho = \text{mass density of air}$$

$$V_H = \text{mean wind speed at height } H$$

The mean wind speed is site dependent and is obtainable as explained in the previous section. In particular, Eq. (5-12) is used to obtain V_H for a specific structural site, as indicated by steps 1 through 4 that immediately follow Eq. (5-14).

The *pressure coefficient* is used to transfer the building-geometry, independent, dynamic wind pressure into a pressure which incorporates, ideally, both site exposure and building geometry. For an example, consider the simplified schematic view of a rigid structure located in a laboratory wind tunnel as shown in Figure 5-11. Imagine that one is able to measure the mean wind speed at point R. Using this measured mean

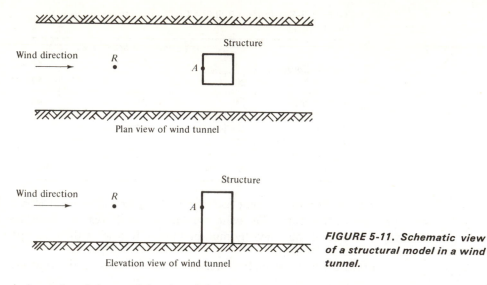

Plan view of wind tunnel

Elevation view of wind tunnel

FIGURE 5-11. Schematic view of a structural model in a wind tunnel.

wind speed and the mass density of the air in the tunnel, one can calculate the dynamic wind pressure at point R by using Eq. (5-15). Next, imagine that the mean wind pressure at point A on the structure, induced by the wind flow, is simultaneously measured by using a pressure tap. The pressure coefficient corresponding to point A is then obtained by dividing the measured mean wind pressure at A by the dynamic wind pressure at point R. In general, each point so located on the structure will have a value for a pressure coefficient at that point. It is therefore possible to obtain a surface area schematic for a structure showing the pressure coefficient for each location. Figure 5-12 shows such a pressure coefficient schematic. These pressure coefficients are referred to as *external pressure coefficients*. Considerable research is being conducted presently to evaluate such coefficients for different structural shapes and site exposures.

The schematic of external pressure coefficients shown in Figure 5-12 is considerably more detailed than those commonly used for the determination of wind loads. Figure 5-13 shows a simplified representation of pressure coefficients. In this figure, one pressure coefficient value is assigned to each of several gross surface-area regions. The direction of the wind for each case is noted (commonly referred to as the *angle of wind attack*). Pressure coefficient values are a function of the angle of attack (ϕ). A positive value of pressure coefficient indicates that the wind pressure pushes on the surface, and a negative value indicates that the wind pressure pulls outward on the surface. The latter is referred to as *suction pressure*. All pressures are assumed to act perpendicular to the surface.

In addition to external pressure coefficients, there are internal pressure coefficients which are a function of the number of openings (e.g., windows or doors) which exist in the structure. Internal pressure coefficients seek to incorporate the wind-flow leakage into the structure and also the difference in ambient air pressure inside and outside the structure due to ambient temperature differentials. The ratio of open surface area to solid surface area for a structure is called the *solidity ratio* (n). For typical multistory buildings this ratio is less than 5%. Table 5-9 shows representative

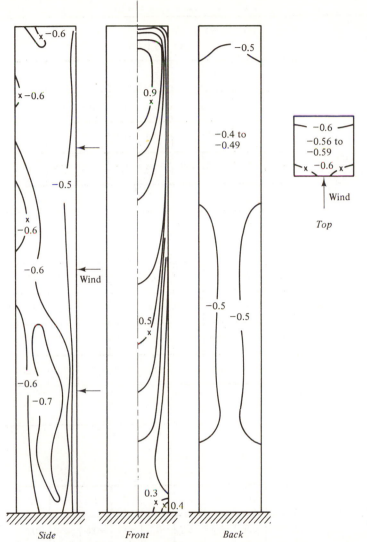

FIGURE 5-12. External pressure coefficients for a tall building model in a wind tunnel [5.4].

values of internal pressure coefficients. The internal pressure coefficient is usually not a function of the internal location of the point under consideration, but is equal for all internal points. The pressure coefficient values of Table 5-9 for uniformly distributed openings have both a plus and minus sign, meaning that both values must be considered for the determination of pressures.

The dynamic wind pressure must be multiplied by the sum of the external and internal pressure coefficients. In some cases, the internal pressure coefficient has no effect on the structural design. For example, as demonstrated later, the total lateral wind force acting on typical structures is not influenced by the values of the internal pressure coefficients.

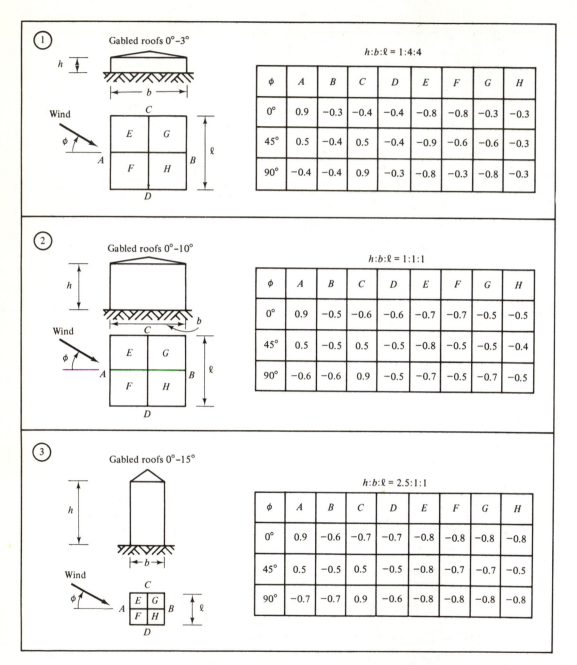

FIGURE 5-13. External pressure coefficients [5.8].

TABLE 5-9. Internal Pressure Coefficients [5.9]

Solidity Ratio	Uniformly Distributed Openings	Openings Mainly In:	
		Windward Wall	*Leeward or Side Walls*
$0 \leq n \leq 0.3$	± 0.3	$+0.3 + 1.67n$	$-0.3 - n$
$n \geq 0.3$	± 0.3	$+0.8$	-0.6

The *gust factor* is incorporated into the determination of a wind pressure so as to include the dynamic variations in the wind speed (i.e., its gustiness) and the vibrational characteristics of the structure. In our previous discussions, the fastest-mile wind speed was used and, as apparent from its definition, that speed is the result of averaging the wind speed over a certain time span (e.g., 1 min for a 60 mph wind). The wind speed varies over this averaging time, and therefore it is not yet included in the pressure calculations. In addition to this source of time variation, the wind flow around the structure introduces additional turbulence. Analytical methods exist for the calculation of the gust factor. These methods include time-series analysis of the wind-speed data and the theory of structural dynamics [5.4]. Therefore, a discussion here is beyond the scope of this book.

The wind pressures acting over the spatial extent of the structure must be integrated so as to determine the total lateral force and moment at any section. The total lateral force and moment at the base of the structure, which must be restrained by foundation design to prevent the structure from sliding and overturning, are usually referred to as the *base shear* and *base overturning moment*, respectively.

Example 5-7

Consider a building located at a site where the wind environment is as described in Example 5-5, part (a).

(a) Calculate the mean dynamic pressure per unit area over the building's height.

(b) If the external pressure coefficients for the building correspond to case 3 in Figure 5-13, calculate the mean wind pressure per unit area on faces *A* and *B*. Assume that $\phi = 0°$.

(c) Calculate the shear and bending moment vs. height for the building. What are the base shear and base overturning moment? Assume uniformly distributed surface openings.

(d) If a gust factor *G* for the building is 1.2, how would your answers for part (c) change?

Part (a) solution:

For the suburban site it follows, from Example 5-5, that

$$V_H = 156.7 \left(\frac{H}{395} \right)^{0.28}$$

and therefore, from Eq. (5-15), the dynamic wind pressure at height H, where H is in meters, is

$$q_D(H) = \frac{1}{2}\rho V_H^2 = \frac{1}{2}(0.00512)\left[156.7\left(\frac{H}{395}\right)^{0.28}\right]^2$$

$$= 2.21H^{0.56} \text{ lb/ft}^2$$

or

$$q_D(H) = 23.8H^{0.56} \text{ lb/m}^2$$

Part (b) solution:

Figure 5-13 shows that on faces A and B the external pressure coefficients are not a function of height and are equal to

$$C_p \text{ at } A = 0.90$$
$$C_p \text{ at } B = -0.60$$

Schematically, this can be shown as indicated in the figure:

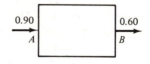

FIGURE Ex. 5-7b

Therefore, the mean wind pressures for faces A and B are

$$q_A(H) = 0.90q_D(H) = 21.4H^{0.56} \text{ lb/m}^2$$
$$q_B(H) = -0.60q_D(H) = -14.3H^{0.56} \text{ lb/m}^2$$

Part (c) solution:

The height-to-width ratio of the building is 2.5 to 1.0 as indicated by Fig. 5-13, and therefore the building width is $(30 \text{ m}/2.5) = 12$ m. The total lateral wind pressure per unit area at height H is

$$q(H) = 21.4H^{0.56} + 14.3H^{0.56}$$
$$= 35.7H^{0.56} \text{ lb/m}^2$$

Note that the internal pressure would not affect the *total* lateral pressure because the internal pressure influence would cancel as indicated in the example figure.

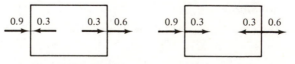

Postive internal pressure   Negative internal pressure   FIGURE Ex. 5-7c

The total lateral wind shear force at a height H above the ground for a building 30 m in height is

$$F(H) = \int_{h=H}^{h=30} (35.7h^{0.56})(12)dh = \frac{(12)(35.7)h^{1.56}}{1.56}\bigg|_{h=H}^{h=30}$$

$$= 55.4 - 0.275H^{1.56} \text{ kips}$$

The shear force at the base of the building (i.e., the base shear) is

$$F(H = 0) = 55.4 \text{ kips}$$

The total lateral wind overturning moment at the height H above the ground for a building 30 m in height is

$$M(H) = \int_{h=H}^{h=30} (35.7h^{0.56})(12)(h - H)dh$$

$$= (35.7)(12)\left[\left(\frac{h^{2.56}}{2.56}\right) - H\left(\frac{h^{1.56}}{1.56}\right)\right]\bigg|_{h=H}^{h=30}$$

$$= 1{,}011{,}672.42 - 129.18H^{15.6} - 167.34H^{2.56}$$

Therefore, the overturning moment at the base of the building (i.e., the base overturning moment) is

$$M(H = 0) = 1{,}011{,}672.42 \text{ kip-m}$$

Part (d) solution:

The total lateral wind pressure per unit area at height H must be multiplied by the gust factor G, and is

$$q(H) = G[35.7H^{0.56}] \text{ lb/m}^2$$

In general, G is a function of building site exposure, height, and the building's vibrational properties. However, if $G = 1.2$, then the total unit force and all resultant calculations are simply scaled by 1.2.

Example 5-8

Consider the building of Example 5-7. Assume that $G = 1.20$ and $n = 0.20$.
(a) Calculate the wind pressure acting on a window on face A at a height of 20 m above the ground.
(b) Repeat part (a), but let the window now be on face B.
(c) A skylight window is located in zone G on the roof. Calculate the wind pressure acting on that window.

Part (a) solution:

The total lateral unit wind pressure acting on a window on face A is

$$q_A(H) = 0.9(1.2)(23.8H^{0.56}) \pm 0.3(1.2)(23.8H^{0.56})$$
$$= 25.70H^{0.56} \pm 8.57H^{0.56} \text{ lb/m}^2$$

Note that the second term has a $+/-$ sign associated with the internal pressures. As the figure in Example 5-7, part (c), indicates the maximum pressure is the sum (i.e., the pushing pressure outside the building and pulling pressure inside the building) and is equal to

$$q_A = 34.27H^{0.56} \text{ lb/m}^2$$

or, at $H = 20$ m,

$$q_A = 183.44 \text{ lb/m}^2 = 17.0 \text{ lb/ft}^2$$

Part (b) solution:

The window pressure on face B is

$$q_B(H) = -0.60(1.2)(23.8H^{0.56}) \pm 0.3(1.2)(23.8H^{0.56})$$

The maximum negative pressure (suction) (i.e., the result of suction from outside and pushing from inside) is

$$q_B(H) = -25.70H^{0.56} \text{ lb/m}^2$$

or, at $H = 20$ m, one obtains

$$q_B(H = 20) = -137.59 \text{ lb/m}^2 = -12.8 \text{ lb/ft}^2$$

Part (c) solution:

The external pressure coefficient in zone G, where the skylight is located, is from Figure 5-13 equal to -0.80. Therefore, the suction internal pressure and the pushing internal pressure combine to produce a maximum negative pressure (suction) equal to

$$q_G = -0.80(1.2)(23.8H^{0.56}) - 0.3(1.2)(23.8H^{0.56}) = -31.42H^{0.56}$$

and, for $H = 30$ m,

$$q_G = -211.03 \text{ lb/m}^2 = -19.61 \text{ lb/ft}^2$$

5-7 LOADING TYPE III

Often, design loads correspond to natural events which occur irregularly in nature. Therefore, unlike the previous loading discussed, one cannot obtain data at predetermined periodic intervals for analysis. Examples of this latter type of loading are earthquakes, tornadoes, and hurricanes. Therefore, one must obtain data when available and analyze that data by using concepts that are different from those explained in the previous sections of this chapter.

In this section we utilize the earthquake problem as a way of developing for the reader some of the basic analysis techniques appropriate for loading type III.

5-8 LOADING TYPE III: DESCRIPTION OF EARTHQUAKE GROUND MOTIONS

The earth's crust has many discontinuities called *faults*. These faults are in a state of stress due to natural forces acting upon them. When the release of this stress state occurs very rapidly an earthquake occurs. The stress release generates seismic waves in the earth, and when such waves pass a site they cause ground-shaking. Buildings and other structures on the ground are subjected to this shaking, and as a result, experience stresses which must be accounted for in design. Earthquake loading exists because it is the structural mass times the earthquake-induced acceleration that is the Newtonian inertia forces which the structure must resist.

Earthquake ground motions and their induced structural motions are not simple natural phenomena that are readily predictable. At the end of this chapter a list of references for further reading is provided. Here we provide a basic introduction to several important aspects of earthquake loading.

Earthquakes occur along a *fault*, which is a discontinuity in the crust of the earth. Such faults exist in many places on this planet. It is believed that the faults are associated with the movement of large plates (see Figure 5-14) within the earth. The study of the movement of these plates is called *plate techtonics* [5.10]. Two kinds of faults are associated with the generation of ground motions which are strong enough to be potentially destructive. The two types are called *strike-slip* and *thrust* faults. Figure 5-15 shows the basic mechanics of the fault rupture for each type.

Two general categories of earthquake waves are induced: they are known as *body waves* and *surface waves*. The body waves consist of compressional waves or longitudinal waves (*P*-waves), and distortional waves or transverse waves (*S*-waves). Surface waves consist of Rayleigh waves and Love waves. The wave with the greatest speed is the *P*-wave (or primary wave). The motion of a unit mass of the earth for this wave type is radial to the source of the energy release (see Figure 5-16), and the speed of this wave is approximately equal to

$$V_p = \text{speed of } P\text{-wave}$$

$$= \sqrt{\frac{(1-v)}{(1+v)(1-2v)}\frac{E}{\rho}} \qquad (5\text{-}16)$$

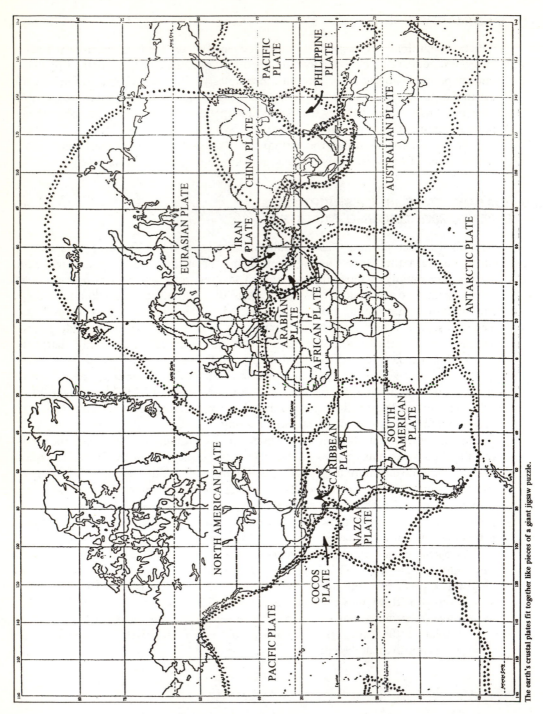

The earth's crustal plates fit together like pieces of a giant jigsaw puzzle.

FIGURE 5-14. World fault plates [5.10].

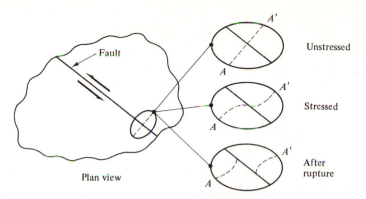

Strike-slip fault and post earthquake rebound

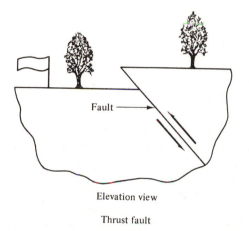

Elevation view

Thrust fault

FIGURE 5-15 Earthquake fault types.

where

$$\rho = \text{mass density of media}$$

$$E = \text{modulus of elasticity of media}$$

$$v = \text{Poisson's ratio of media}$$

The S-wave (or shear or secondary wave) is the second fastest wave, and the motion of a unit mass of the earth is transverse to the lines from the source of energy release to the unit mass (see Figure 5-16). The speed of the S-wave is approximately equal to

$$V_S = \text{speed of } S\text{-wave} = \sqrt{\frac{G}{\rho}} \qquad (5\text{-}17)$$

where

$$G = \text{shear modulus of media} = \frac{E}{2(1 + v)} \qquad (5\text{-}18)$$

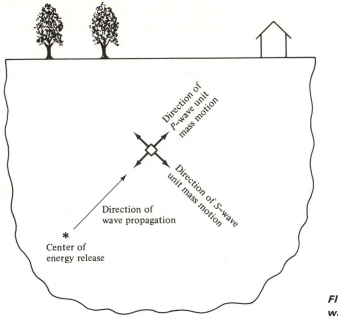

FIGURE 5-16. P-waves and S-waves.

The surface wave travels along the surface of the earth. The *P*-wave and *S*-wave travel through and reflect off the surface of the earth and, for that reason, are often referred to as *body waves*. The surface wave travels at a speed approximately equal to 90% of the *S*-wave speed. Seismologists use the elapsed time between the arrival of a *P*-wave and an *S*-wave at a given site to assist them in estimating the distance from the site to the center of energy release. The *P*-wave speed in rock near the surface of the earth is approximately 6 to 7 km/sec. The corresponding *S*-wave speed in rock is approximately 4 km/sec.

Figure 5-17 shows a simplified cross-sectional view of a portion of the earth. The following terms are defined for this figure:

Focus (or *hypocenter*): The center of the initial rupture by an earthquake.

Epicenter: A point on the earth's surface directly above the focus.

Focal depth: The depth of the focus beneath the surface.

Two other terms which depend upon the specification of a particular site (e.g., a dam site or a building site of interest) are:

Epicentral distance: The distance from the epicenter to the site.

Hypocentral distance: The distance from the focus or hypocenter to the site.

Perhaps the most familiar term used to characterize an earthquake is its *Richter magnitude*. This magnitude is based on an experimental reading obtained on an instrument called a *Wood-Anderson seismograph*, at a specified distance of 100 km from the

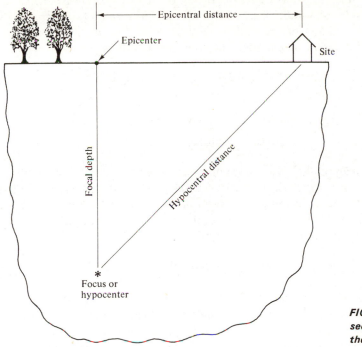

FIGURE 5-17. Schematic cross-sectional view of a portion of the earth.

epicenter of an earthquake. If an instrument is not available at this distance from the epicenter, then empirical formulas are used to obtain an estimate of the 100 km reading. The Richter magnitude is obtained for a given earthquake by using the following formula [5.11]:

$$M = \text{Richter magnitude} = \log \left(\frac{A_{100}}{A_R} \right) \tag{5-19}$$

where

A_{100} = maximum amplitude measured for the earthquake on the Wood-Anderson seismograph at a disance of 100 km from the epicenter

A_R = the reference amplitude level of a specific Wood-Anderson seismograph

The Richter magnitude is a measure of the energy released by an earthquake; an estimate of that energy is obtained by using

$$\log W = 11.8 + 1.5M \tag{5-20}$$

where

$$W = \text{energy (ergs)}$$

The Richter magnitude has been empirically related to the relative displacement between the two sides of a fault at ground surface and also the total length of a fault rupture. These two parameters provide a physical indication of the relative severity of earthquakes of different magnitude. The empirical relations are

$$\log D = 0.57M - 3.39 = 0.86 \log L - 0.46 \tag{5-21}$$

where

$$\begin{aligned} D &= \text{relative fault displacement (ft)} \\ L &= \text{length of fault rupture (miles)} \end{aligned} \tag{5-22}$$

The Richter magnitude is a measure of the surface wave amplitude based on the Wood-Anderson seismograph instrument characteristics and the reference distance of 100 km. An alternative magnitude has been proposed, called the *body wave magnitude (m)*, which is more directly a measure of *P*- and *S*-wave amplitudes. The two magnitudes are commonly related by using the formula

$$M = 1.59m - 3.97 \tag{5-23}$$

The *Modified Mercalli Intensity* (MMI) is a parameter used to describe the observed effect of ground-shaking at a particular site. Table 5-10 defines the MMI scale. The MMI value assigned after an earthquake at a particular site is a subjective evaluation of damage made by an observer and is particularly valuable in lieu of instrumental records of ground motion. An empirical relationship between MMI and peak ground velocity is as follows [5.12]:

$$\text{MMI} = \frac{\log 14v}{\log 2} \tag{5-24}$$

where

$$v = \text{peak ground velocity (cm/sec)}$$

In recent years with the technological advances made in instrumentation and their reduction in cost, the acquisition of time histories of ground motion has become more common. An instrument called a *strong motion accelerograph* is now placed on the ground to measure earthquake-induced motion. The three orthogonal components of ground acceleration are measured by using this instrument. Figure 5-18 shows a copy of the records obtained by an accelerograph. Each acceleration vs. time trace is called an *accelerogram*. Figure 5-19 shows a representative accelerogram. The earthquake accelerogram can be directly interpreted to provide estimates of peak ground acceleration, duration of strong ground-shaking, and frequency content. The acceleration vs. time record can be integrated to obtain time histories of ground velocity and ground displacement.

The accelerogram is used to calculate an extremely important structural engineering quantity called a *response spectrum*, which is a plot, vs. the natural period or

TABLE 5-10. Modified Mercalli Intensity Scale [5.12]

MMI

    I. Not felt except under especially favorable circumstances.

   II. Felt by persons at rest, on upper floors, or favorably placed.

  III. Felt indoors. Hanging objects swing. Vibration like passing of light trucks. May not be recognized as an earthquake.

  IV. Hanging objects swing. Vibration like passing of heavy trucks or sensation of a jolt like a heavy ball striking the walls. Standing motor cars rock. Windows, dishes, doors rattle. Glasses clink. Crockery clashes. Wooden walls and frames creak.

   V. Felt outdoors; direction estimated. Sleepers wakened. Liquids disturbed; some spilled. Small unstable objects displaced or upset. Doors swing, close, open. Shutters, pictures move.

  VI. Felt by all. Persons walk unsteadily. Windows, dishes, glassware broken. Knickknacks, books, etc., off shelves. Pictures off walls. Furniture moved or overturned. Weak plaster and masonry D cracked. Small bells ring (church, school). Trees, bushes shaken visibly or heard to rustle.

 VII. Difficult to stand. Noticed by drivers of automobiles. Hanging objects quiver. Furniture broken. Damage to masonry D, including cracks. Weak chimneys broken at roofline. Fall of plaster, loose bricks, stones, tiles, cornices; also unbraced parapets and architectureal ornaments. Some cracks in masonry C. Waves on ponds; water turbid with mud. Small slides and caving in along sand or gravel banks. Large bells ring. Concrete irrigation ditches damaged.

VIII. Steering of automobiles affected. Damage to masonry C; partial collapse. Some damage to masonry B; none to masonry A. Fall of stucco and some masonry walls. Twisting, fall of chimneys, factory stacks, monuments, towers, elevated tanks. Frame houses moved on foundations if not bolted down; loose panel walls thrown out. Decayed piling broken off. Branches broken from trees. Changes in floor or temperature of springs and wells. Cracks in wet ground and on steep slopes.

  IX. General panic. Masonry D destroyed; masonry C heavily damaged, sometimes with complete collapse; masonry B seriously damaged. General damage to foundations. Frame structures, if not bolted, shifted off foundations. Frames racked. Serious damage to reservoirs. Underground pipes broken. Conspicuous cracks in ground. In alluviated area, sand and mud ejected, earthquake foundations, and craters.

   X. Most masonry and frame structures destroyed with their foundations. Some well-built wooden structures and bridges destroyed. Serious damage to dams, dikes, embankments. Large landslides. Water thrown on banks of canals, rivers, lakes, etc. Sand and mud shifted horizontally on beaches and flat land. Rails bent slightly.

  XI. Rails bent greatly. Underground pipelines completely out of service.

 XII. Damage nearly total. Large rock masses displaced. Lines of sight and level distorted. Objects thrown into the air.

Definition of Masonry A, B, C, D:

    Masonry A: Good workmanship, mortar, and design; reinforced, especially laterally and bound together by using steel, concrete, etc.; designed to resist lateral forces.

    Masonry B: Good workmanship and mortar; reinforced but not designed in detail to resist lateral forces.

    Masonry C: Ordinary workmanship and mortar; no extreme weaknesses like failing to tie in at corners, but neither reinforced nor designed against horizontal forces.

    Masonry D: Weak materials such as adobe; poor mortar; low standards of workmanship, weak horizontally.

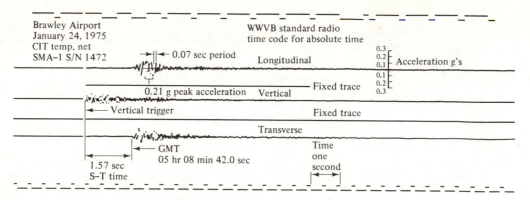

FIGURE 5-18. Typical strong motion earthquake accelerogram [5.13].

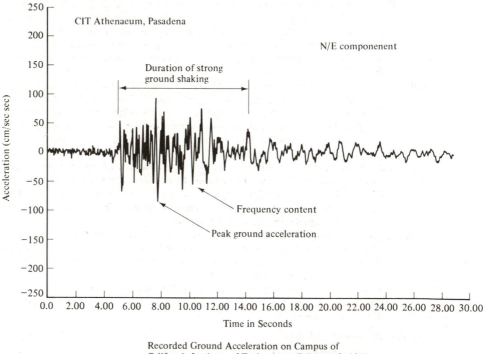

Recorded Ground Acceleration on Campus of
California Institute of Technology, February 9, 1971

FIGURE 5-19. Typical strong motion earthquake accelerogram [5.13].

frequency of vibration of a single-degree-of-freedom oscillator, of the peak oscillator response when subjected to a base excitation equal to the earthquake accelerogram. The plot is for a specified value of oscillator damping. The response most often calculated is the oscillator relative displacement. For a specified oscillator natural period of vibration (T_n) and damping (ξ) the maximum relative displacement is called the *spectral displacement* and is denoted as $S_d(T_n, \xi)$. The plot of $S_d(T_n, \xi)$ vs. T_n indicates the relative amplitude of motion which oscillators with different T_n values can expect. Two additional parameters are now defined, which are directly calculated from the spectral displacement. They are

$$S_v(T_n, \xi) = \text{spectral velocity}$$

$$= \left(\frac{2\pi}{T_n}\right) S_d(T_n, \xi) \tag{5-25}$$

and

$$S_a(T_n, \xi) = \text{spectral acceleration}$$

$$= \left(\frac{2\pi}{T_n}\right)^2 S_d(T_n, \xi) \tag{5-26}$$

These defined parameters also indicate the amplitude of response of the oscillator for different T_n values. In addition, the spectral velocity for zero percent oscillator damping is an upper-bound estimate of the accelerogram's Fourier modulus. The spectral acceleration is a good approximation, for damping less than 15%, of the maximum absolute acceleration of the oscillator mass. Figure 5-20 shows spectral velocity plots in two forms for a representative earthquake. Part (a) of the figure has a linear scale on each axis. Part (b) is referred to as the *three-way log plot*. The vertical axis in a three-way log plot is the spectral velocity, and the axes inclined at 45° to the left or right are used to obtain the magnitude of spectral acceleration or spectral displacement, respectively. The maximum ground motion displacement, velocity, and acceleration are commonly noted on the same plot, as indicated, so as to clearly indicate the amplification or attenuation characteristics of the oscillator response relative to the peak ground motion.

Engineers are now beginning to reliably relate the values of peak ground motion or spectral response parameters to Richter magnitude and hypocentral distance. This development of relationships is now possible because of the increased number of earthquake accelerograms. These parametric relationships are referred to as *attenuation relationships*. Consider as an example the relationship between peak ground acceleration and Richter magnitude and hypocentral distance. Among the many alternative expressions available in the research literature consider the following two for illustrative purposes [5.14, 5.15]:

$$a = 1230e^{0.80M}(R + 25)^{-2} \tag{5-27}$$

and

$$a = b_1 10^{b_2 M}(R + 25)^{-b_3} \tag{5-28}$$

where

$$a = \text{peak ground acceleration (cm/sec}^2)$$
$$R = \text{hypocentral distance (km)}$$

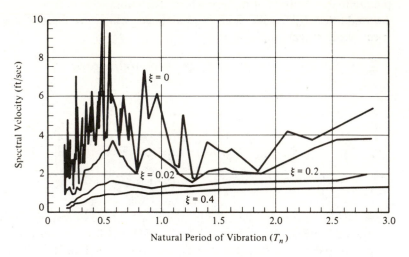

(a) Standard linear/linear plot

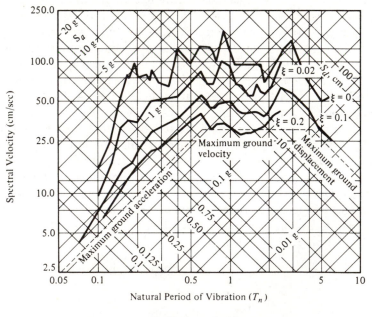

(b) Three way log plot

FIGURE 5-20. Two types of spectral velocity plots.

$$b_1 = 472.3$$
$$b_2 = 0.278$$
$$b_3 = 1.301$$

These alternative formulas estimate different peak acceleration values for a given R and M. The difference is typical of that obtained by using other alternative deterministic ground acceleration attenuation relationships.

When any ground motion or spectral response parameter (e.g., peak ground acceleration, S_v or S_a) is considered to be a random variable, then one needs to proceed with a characterization of the type discussed in Chapter 2. Consider as an example the characterizations shown in Figures 5-21 and 5-22 [5.16]. In Figure 5-21 the peak ground acceleration is the random variable, and its mean and mean plus and minus 1 and 2 standard deviations is related to hypocentral distance for Richter magnitude 6.5 earthquakes. Figure 5-22 shows a plot of the mean and mean plus 1 standard deviation when the spectral acceleration divided by peak ground acceleration is considered to be a random variable.

Table 5-11 indicates a proposed alternative form for the characterization of such

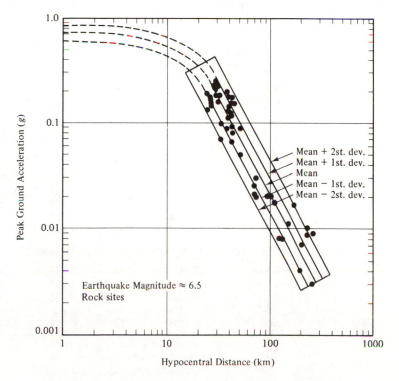

FIGURE 5-21. Peak ground acceleration uncertainty [5.16].

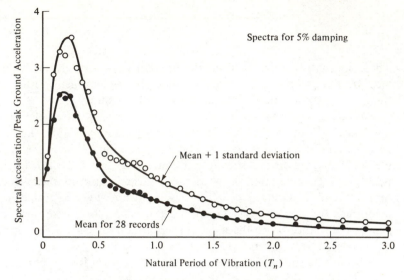

FIGURE 5-22 Spectral acceleration uncertainty [5.16].

TABLE 5-11. Attenuation Uncertainty Characterization [5.15]

y	b_1	b_2	b_3	Coefficient of Variation of y
a gal	472.3	0.278	1.301	0.548
v cm/sec	5.64	0.401	1.202	0.969
d cm	0.393	0.434	0.885	0.883
Spectral Velocity (0% damping) S_v (cm/sec)				
$T_n = 0.1$ sec	11.0	0.278	1.346	0.941
0.5	3.05	0.391	1.001	0.636
1.0	0.631	0.378	0.549	0.768
2.0	0.9768	0.469	0.419	0.989
5.0	0.0834	0.564	0.897	1.344
Spectral Velocity (5% damping) S_v (cm/sec)				
$T_n = 0.1$ sec	10.09	0.233	1.341	0.651
0.5	5.74	0.356	1.197	0.591
1.0	0.432	0.399	0.704	0.703
2.0	0.122	0.466	0.675	0.941
5.0	0.0706	0.557	0.938	1.193

a = peak acceleration of ground (gal)
v = peak velocity of ground (cm/sec)
d = peak displacement of ground (cm)
1 gal = 1 cm/sec^2

random variables where the equation [5.15]

$$y = b_1 10^{b_2 M}(R + 25)^{-b_3} \qquad (5-29)$$

is used for several variables. The column at the extreme left of the table denotes the random parameter, and the column at the extreme right denotes the coefficient of variation of the random variable. For example, Eq. (5-28) is used to calculate the mean value of the peak ground acceleration, and the coefficient of variation of this random variable is 0.548. This latter uncertainty characterization differs from that shown in Figures 5-21 and 5-22 in that the former incorporates a characterization of the site's soil characteristics.

The preceding attenuation relationships considered an earthquake's Richter magnitude and hypocentral distances to be given information. Consider now the uncertainty as to the actual occurrence of an earthquake and, in particular, its Richter magnitude and location relative to a site in question. This area of study is referred to as *seismicity*. In a seismicity study the first step is the acquisition of historical information on the Richter magnitudes and locations of past earthquakes in the vicinity of the site. Consider as an illustrative example the southern California region of the United States. Table 5-12 summarizes all earthquakes of Richter magnitude 6 or greater during the period of 1912 to 1974 [5.17]. Figure 5-23 shows the location of these earthquake epicenters. Thirty-four earthquakes have occurred during this 63-year time span. The data in Table 5-12 can be analyzed by using the techniques explained in Chapter 2. That is, the 34 values of Richter magnitude can be visualized as 34 data points from which statistical moments, histograms, and probability density

TABLE 5-12. Southern California Earthquakes of Magnitude 6 or Greater (1912–1974) [5.17]

Year	Richter Magnitude	Year	Richter Magnitude
1915	6.25	1946	6.3
	6.25	1947	6.2
	7.1	1948	6.5
1916	6.0	1952	7.7
1918	6.8		6.4
1923	6.25		6.1
1925	6.3		6.1
1927	6.0	1954	6.2
1933	6.3		6.3
1934	6.5		6.0
	7.1	1956	6.8
1935	6.0		6.1
1937	6.0		6.3
1940	6.7		6.4
	6.0	1966	6.3
1941	6.0	1968	6.4
1942	6.5	1971	6.4

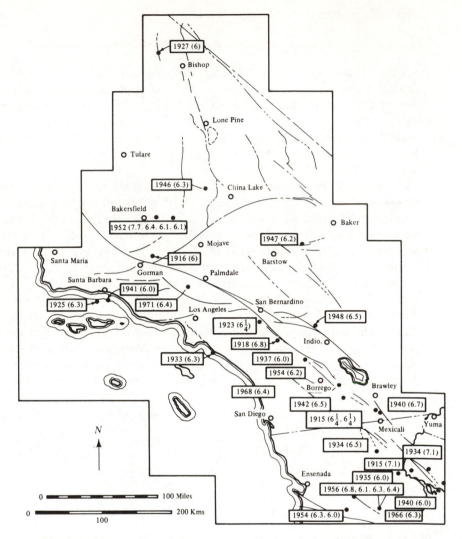

FIGURE 5-23. Southern California earthquakes of magnitude 6 or greater (1912–1975) [5.17]. ("An Evaluation of a Response Spectrum Approach to Seismic Design of Buildings," Applied Technology Council, 1974.)

functions follow. For example, the mean Richter magnitude of this data is 6.37. Table 5-13 shows a tabular summary of the data by Richter magnitude intervals, and Figure 5-24 shows the cumulative histogram of the data. When the data is analyzed in this manner it is considered as a set of data for a time span of 63 years. An extension of the concepts presented in Chapter 2 is necessary to more directly incorporate the time variable. The data can be normalized by dividing the number of earthquakes equal to or less than a certain Richter magnitude by the number of years of data. Alternatively, and more commonly, one can divide the number of events equal to or

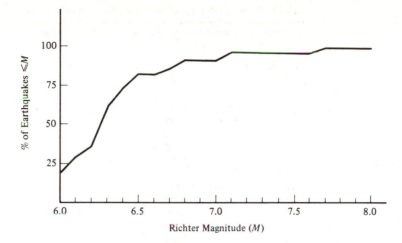

FIGURE 5-24. *Cumulative frequency histogram for Southern California earthquake data (1912–1974).*

TABLE 5-13. Summary of Southern California Earthquake Data

Richter Magnitude Interval	Number of Earthquakes	Number of Earthquakes with Magnitude ≤ Upper interval limit
6.0 –6.10	10	10 (29%)
6.11–6.20	2	12 (35%)
6.21–6.30	9	21 (62%)
6.31–6.40	4	25 (74%)
6.41–6.50	3	28 (82%)
6.51–6.60	0	28 (82%)
6.61–6.70	1	29 (85%)
6.71–6.80	2	31 (91%)
6.81–6.90	0	31 (91%)
6.91–7.00	0	31 (91%)
7.01–7.10	2	33 (97%)
7.11–7.20	0	33 (97%)
7.21–7.30	0	33 (97%)
7.31–7.40	0	33 (97%)
7.41–7.50	0	33 (97%)
7.51–7.60	0	33 (97%)
7.61–7.70	1	34 (100%)
Above 7.70	0	34 (100%)

greater than a certain Richter magnitude by the number of years of data. Table 5-14 shows a summary in this latter form, where N is defined as the number of earthquakes per year equal to or greater than the specified Richter magnitude. The inverse of the number of events per year is the *return period.* Note that this definition of return period is basically the same as was used for discussing wind-speed return period. For example, there are three earthquakes with a Richter magnitude 7.1 or greater during the 63

TABLE 5-14. Earthquakes per Year and Return Periods

Richter Magnitude	Number of Earthquakes per Year $\geq$ M	Return Period (years) (T = 1/N)
6.0	0.540	1.9
6.1	0.429	2.3
6.2	0.381	2.6
6.3	0.302	3.3
6.4	0.206	4.9
6.5	0.143	7.0
6.6	0.095	10.5
6.7	0.095	10.5
6.8	0.079	12.7
6.9	0.048	21
7.0	0.048	21
7.1	0.048	21
7.2	0.016	63
7.3	0.016	63
7.4	0.016	63
7.5	0.016	63
7.6	0.016	63
7.7	0.016	63

years of data, and therefore the number of earthquakes per year with a Richter magnitude equal to or greater than 7.1 is equal to $N = 3/63$. The corresponding return period (T) is defined as the inverse of N, or $T = 1/N = 63/3 = 21$ years.

It is often desirable to fit an equation to this type of data, and the equation most commonly used is

$$\log N = a - bM \tag{5-30}$$

where a and b are constants determined by a least-squares fit of the data. For the data presented in Table 5-14 the equation is

$$\log N = 5.76 - 1.01M \tag{5-31}$$

One reason for developing an equation relating N and M is that one can extrapolate beyond the data and determine return periods for earthquakes of larger Richter magnitude. Such a practice can be dangerous and is not desirable. However, since engineers are required to prepare design calculations only by using the data available at the time of design, this practice is often unavoidable.

A return period estimate provides one with a very helpful measure of earthquake seismicity. However, it only indicates that on the average one can expect an earthquake of a specified Richter magnitude or greater to occur every T years. This description does not go far enough, and a probability must be assigned as to the future occurrence of one or more earthquakes of a certain Richter magnitude, or greater, within a future time span. The most common method of doing this is to utilize the Poisson probability density function. This function is defined as

$$p(n) = \frac{(t/T)^n e^{-(t/T)}}{n!} \qquad n = 0, 1, 2, \ldots \tag{5-32}$$

where

$T =$ return period in years

$p(n) =$ probability of having exactly n occurrences during the next t years

The probability of *nonoccurrence* is a commonly referred to parameter, and it is obtained using Eq. (5-32) with $n = 0$. Another common parameter is the *probability of occurrence*, and it is defined as 1 minus the probability of nonoccurrence; i.e.,

$$p_0 = \text{probability of occurrence} = 1 - e^{-(t/T)} \tag{5-33}$$

Note that the probability of occurrence within a time span equal to the return period is only 63.2%.

The Poisson PDF is most commonly used in earthquake engineering. However, the reader should recognize that it can also be used for type II loads (e.g., wind), wherein the return period then corresponds to that previously discussed in the wind section of this chapter.

Earthquakes occur with different return periods in different geographic regions. Figure 5-25 shows the occurrence of earthquakes in most states. However, the intensity of ground-shaking is markedly different between geographic regions. *Seismic zoning* is concerned with the division of geographic areas into regions, or zones, of approximately equal earthquake activity. The method used to identify and characterize these zones is not universal. Figure 5-26 shows zoning maps for the continental United States on a 0 to 4 scale, with 4 being the most active. Seismic zoning maps are often referred to as *seismic risk maps*. Seismic zoning maps which relate ground motion parameters to return periods are based on a statistical analysis of the data and incorporate many of the items previously discussed in this chapter. Figure 5-27 shows such a map for the southern California area and a 100 year return period.

Independent of methodology, seismic zoning procedures must incorporate historic earthquake activity, fault locations, and soil characteristics. The development of improved procedures for estimating earthquake ground motion is one of the most active areas of earthquake engineering research. A geotechnical engineer is one who specializes in this area, and this type of engineer should be consulted to obtain the most recent professional input on this topic. The basic concepts involved in seismicity studies must be understood by the structural engineer because he must use the results of such studies and therefore must ensure that the information provided is meaningful for structural design.

Earthquakes have occurred in nearly all of the 50 states.

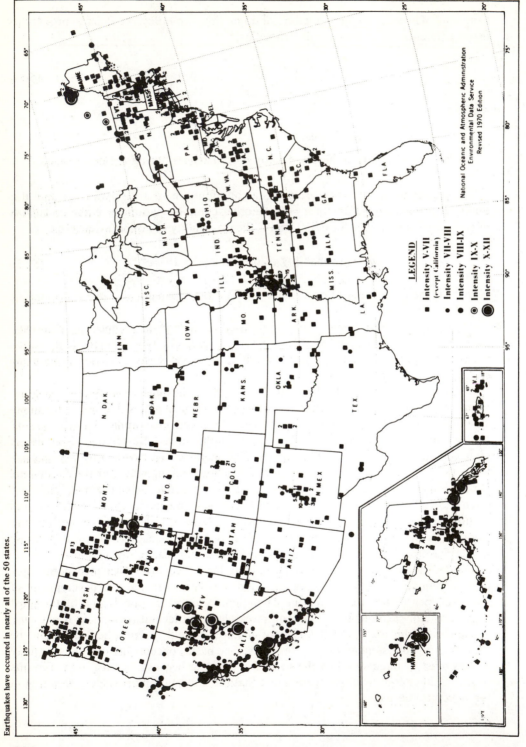

FIGURE 5-25. U.S. earthquake activity [5.18].

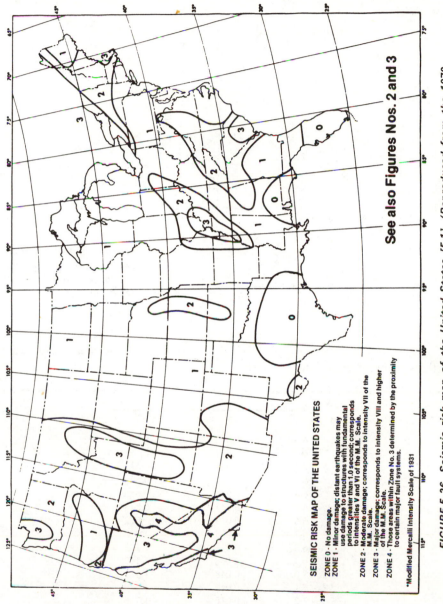

SEISMIC RISK MAP OF THE UNITED STATES

ZONE 0 - No damage.
ZONE 1 - Minor damage; distant earthquakes may
use damage to structures with fundamental
periods greater than 1.0 second; corresponds
to intensities V and VI of the M.M. Scale.

ZONE 2 - Moderate damage; corresponds to intensity VII of the
M.M. Scale.

ZONE 3 - Major damage; corresponds to intensity VIII and higher
of the M.M. Scale.

ZONE 4 - Those areas within Zone No. 3 determined by the proximity
to certain major fault systems.

*Modified Mercalli Intensity Scale of 1931

See also Figures Nos. 2 and 3

FIGURE 5-26. Seismic risk map of the United States [5.1]. (Reproduced from the 1979 edition of the Uniform Building Code, copyright 1979, with permission of the publisher, the International Conference of Building Officials.)

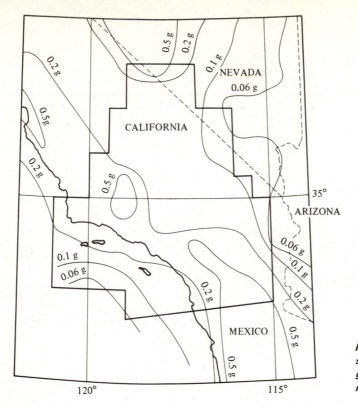

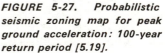

FIGURE 5-27. *Probabilistic seismic zoning map for peak ground acceleration: 100-year return period [5.19].*

Example 5-9

Calculate the coefficients a and b in the equation $\log N = a + bM$, using the earthquake occurrence data in Table 5-14.

M_i	N_i	$\log N_i$
6.0	0.540	−0.268
6.1	0.429	0.368
6.2	0.381	−0.419
6.3	0.302	−0.520
6.4	0.206	−0.686
6.5	0.143	−0.845
6.6	0.095	−1.022
6.7	0.095	−1.022
6.8	0.079	−1.102
6.9	0.068	−1.167
7.0	0.048	−1.319
7.1	0.048	−1.319
7.2	0.016	−1.796
7.3	0.016	−1.796
7.4	0.016	−1.796
7.5	0.016	−1.796
7.6	0.016	−1.796
7.7	0.016	−1.796

It follows that

$$\bar{M} \equiv \frac{1}{18} \sum M_i = 6.85$$

$$\overline{\log N} \equiv \frac{1}{18} \sum \log N_i = -1.157$$

and therefore

$$b = \frac{\sum M_i(\log N_i) - n(\bar{M})(\overline{\log N})}{\sum M^2 - n(\bar{M})^2} = \frac{-147.634 - 18(6.85)(-1.157)}{849.45 - 18(6.85)^2} = -1.03$$

and therefore using a least squares fit

$$a = (\overline{\log N}) - b(\bar{M}) = -1.157 - (-1.03)(6.85) = 5.90$$

Finally,

$$\log N = 5.90 - 1.01M$$

Example 5-10

Figure Ex. 5-10 summarizes the number of earthquakes per year, throughout the world, equal to or greater than certain Richter magnitude levels. The equation

$$\log N = 7.7 - 0.9M$$

was used to fit the data over the Richter magnitude range 6.0 to 8.0, with the dashed curve applying above Richter magnitude 8.0.

(a) Calculate the return period of an earthquake of Richter magnitude 6.0 or greater.
(b) What is the probability of having n earthquakes of Richter magnitude 8 or greater in the next year?
(c) What is the return period of an earthquake having a Richter magnitude between 7 and 8?

Part (a) solution:

From the Equation $\log N = 7.7 - 0.9M$ for $M = 6$ or greater, $N = 2 \times 10^2$ and $T = 1/N = 0.005$ year (i.e., almost daily!). Similarly, for $M = 8$ or greater, $N = 3.16$ and $T = 0.316$ year.

Part (b) solution:

The Poisson PDF is

$$p(n) = \frac{(t/T)^n e^{-(t/T)}}{n!}$$

and, in this case,

$$T = 0.316 \text{ year}$$
$$t = 1.00 \text{ year}$$

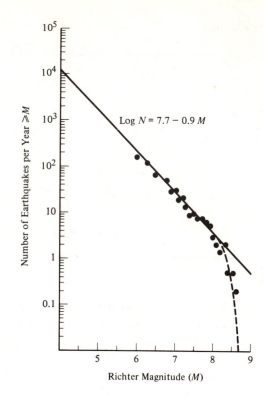

Number of Earthquakes per Year ≥ M

$$\text{Log } N = 7.7 - 0.9\,M$$

Richter Magnitude (M)

FIGURE Ex. 5-10

Therefore,

$$p(n) = \frac{(3.16)^n e^{-3.16}}{n!} = \frac{0.041(3.16)^n}{n!}$$

or

n	p(n)	
0	0.041	
1	0.134	
2	0.212	
3	0.223	Most probable
4	0.177	number of earthquakes
5	0.112	
6	0.059	
7	0.027	
8	0.011	
9	0.004	
10	0.001	

Part (c) solution:

The number of earthquakes per year with a Richter magnitude 7 or greater is 25.12, and the number per year with a Richter magnitude 8 or greater is 3.16. Therefore, the number per year with a Richter magnitude greater than or equal to 7 but less than 8 is $25.12 - 3.16 = 21.96$. Thus, the return period is $T = (1/21.96)$ or 0.046 years or 2.39 weeks.

Example 5-11

Consider a building site: in the region surrounding the site nine earthquakes of magnitude 6 or greater have been observed during a 100 year period. The hypocentral distance to the site for each earthquake was determined and was as noted in the following table. An attenuation formula was used to calculate the peak ground acceleration for each earthquake and is also given in the table.

(a) Determine the number of earthquakes per year with ground accelerations equal to or greater than $0.05g$, $0.10g$, $0.15g$, ..., $0.35g$, at the site.

(b) Calculate the return period of ground accelerations equal to or greater than each of the acceleration levels from a log N vs. a plot for this site.

(c) Calculate the return period for peak ground accelerations equal to or greater than $0.20g$ but less than $0.30g$. Use the straight line from part (b).

(d) Calculate the probability of having exactly one earthquake with a ground acceleration equal to or greater than $0.03\,g$ at the site in the next 5 years.

Richter Magnitude (M)	Hypocentral Distance (km) (R)	Peak Ground Acceleration at Site (g) (a)
7	6.7	0.337
6.6	1.2	0.359
6.1	3.3	0.206
6.7	8.3	0.241
6.8	3.3	0.361
6.5	3.67	0.060
6.4	21.7	0.096
8.25	70.	0.102
8.0	71.7	0.081

Part (a) solution:

Using the above table the number of accelerations per year are as follows:

a	N	Return Period (years) Using (1/N)	Using Curve
0.05g	0.09	11.1	11.1
0.10	0.06	16.7	14.3
0.15	0.05	20.0	18.5
0.20	0.05	20.0	23.3
0.25	0.03	33.3	30.3
0.30	0.03	33.3	38.5
0.35	0.02	50.0	50.0

Part (b) solution:

The plot shown in Fig. Ex. 5-11 is obtained by passing a straight line through the data. The return periods in the right-hand column in the table above result. Note how they compare with the return periods obtained using the raw data.

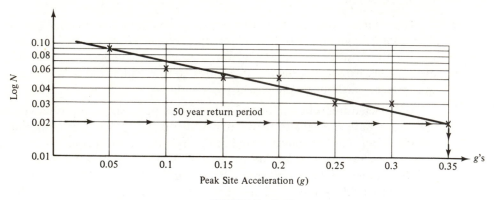

FIGURE Ex. 5-11

Part (c) solution:

For the value $a = 0.20g$, $N = 0.043$, and for $a = 0.30g$, $N = 0.026$. Therefore, the number of peak ground accelerations per year with $0.20 \leq a < 0.30g$ is $N = 0.043 - 0.026 = 0.0170$, and the return period is $T = 1/N = 1/0.0170 = 58.8$ years.

Part (d) solution:

The Poisson PDF is

$$p(n) = \frac{(t/T)^n e^{-(t/T)}}{n!}$$

and, in this example, $n = 1$, $t = 5$ years, $T = 38.5$ years. Therefore,

$$p(n = 1) = \frac{5}{38.5} e^{-5/38.5} = 0.114, \text{ or } 11.4\%$$

5-9 LOADING TYPE III: DESCRIPTION OF EARTHQUAKE LOADS

A structure when subjected to an earthquake movement at the foundation level responds by moving. Such movement results in an acceleration of all component parts of the structure, and it is that acceleration times the particle (or component) mass which results in Newton inertia forces (or loads). These inertia forces produce strains and stresses in the structure. The engineer must ensure that the structure is designed such that these load-induced stresses are not excessive to the point of causing a collapse or a major economic loss.

The earthquake design of a structure is a very complex problem. The ground provides a dynamic input which changes in a most irregular manner as a function of time (see Figure 5-19). Until recently, structural engineering students typically had little formal education in the area of structural dynamics. Therefore, most structural engineers have used equivalent static earthquake loads to calculate earthquake-induced stresses—the basic concept being the substitution of simply derived static loading for use in earthquake design instead of more sophisticated and exact loading which results from a dynamic analysis. In recent years, the response spectrum discussed in the previous section has become accepted by an increasing number of engineers as the more appropriate earthquake input for design. The response spectrum earthquake design approach which uses this input is slowly replacing the equivalent static earthquake design approach for most major structures such as buildings and hospitals.

An equivalent static earthquake design approach can be found in most national building codes. The earthquake loading is a function of the historical earthquake activity or seismicity of the general region where the structure is to be built, the type of structure, and the basic dynamic characteristics of the structure. The Uniform Building Code (UBC) also expresses the loads as a function of the type of occupancy and the soil characteristics at the site of the structure.

The equivalent static approach can be visualized as being composed of two basic parts. First, one calculates the total design lateral earthquake force. Second, one distributes this force over the spatial extent of the structure. In both of these parts the earthquake-induced force is not considered to be a function of time.

Consider the schematic representation of a building as shown in Figure 5-28. Denoting the total dead weight of the building by the symbol W and the total lateral earthquake-induced force by the symbol V, the 1979 Uniform Building Code relates the total lateral earthquake force to be building dead weight by the formula

$$\frac{V}{W} = ISZKC \qquad (5\text{-}34)$$

where

$I =$ building occupancy importance factor (1 to 1.5)

$S =$ site-structure resonance factor (1 to 1.5)

$Z =$ seismicity zone factor (0 to 1)

$K =$ lateral structural force-resistant system factor (0.67 to 1.33)

$C = \frac{1}{15}\sqrt{T_n} \le 0.12$ and $C = \dfrac{1}{15\sqrt{T_n}}$

$T_n =$ fundamental elastic period of vibration of the building (in seconds) in the direction under consideration

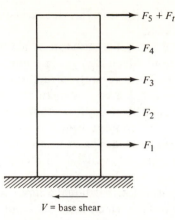

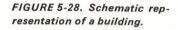

V = base shear

FIGURE 5-28. Schematic representation of a building.

The occupancy importance factor is 1.5 for essential building facilities (e.g., schools, fire stations, hospitals); 1.25 for any building where the primary occupancy is for the assembly of more than 300 persons in one room; and 1.0 for all other buildings.

The site-structure resonance factor is equal to

$$S = 1.0 + \left(\frac{T_n}{T_s}\right) - 0.5\left(\frac{T_n}{T_s}\right)^2 \text{ for } \frac{T_n}{T_s} \leq 1 \tag{5-35a}$$

or

$$S = 1.2 + 0.6\left(\frac{T_n}{T_s}\right) - 0.3\left(\frac{T_n}{T_s}\right)^2 \text{ for } \frac{T_n}{T_s} > 1 \tag{5-35b}$$

where

T_s = characteristic fundamental natural period of vibration of the soil at the site (in seconds)

The seismic zone factor is dependent upon the location of the building site. The value of Z is dependent upon the site location as related to the zones shown in Figure 5-26. For zones 0, 1, 2, 3, and 4 of Figure 5-26, the Z values are 0, 3/16, 3/8, 3/4, and 1, respectively.

The total lateral earthquake load is a function of the type of structural system used in the building design. The lowest value of K corresponds to a building with a ductile moment-resisting steel space frame. A concrete shear wall building would have a K of 1.33.

The total lateral earthquake-induced force V is most commonly referred to as the *lateral base shear force*. This force is distributed vertically over the building, using the formula

$$V = F_t + \sum_{i=1}^{n} F_i \tag{5-36}$$

where

n = number of floors or stories in the building

F_t = that portion of V applied at the top floor in addition to F_i for $i = n$

F_i = that portion of V applied at the ith floor of the building

It is also given that the supplemental force at the top floor is

$$F_t = 0.07T_nV \leq 0.25W \qquad (5\text{-}37a)$$

and

$$F_t = 0 \text{ if } T_n \leq 0.7 \text{ sec} \qquad (5\text{-}37b)$$

The lateral earthquake force at each floor is calculated by using

$$F_i = \frac{(V - F_t)W_ih_i}{\sum\limits_{x=1}^{n} W_xh_x} \qquad (5\text{-}38)$$

where

w_x, w_i = dead weight of the xth, ith floor

h_x, h_i = height (in feet) above the base of the building to xth, ith floor

It is important to observe the basic approach presented by the equivalent static loading procedure because the numbers have, and will continue, to be changed in codes but the approach will in all probability be the same for some time.

The total lateral earthquake force divided by the dead weight of the structure is a function of T_n, the fundamental natural period of building vibration. The values of the other factors (e.g., the occupancy factor and framing-type factor) are established by technical committees composed of structural engineers and are based on experience and technical research. However, the approach of quantifying a V/W ratio has its roots in a response spectral characterization of earthquake ground motion. To illustrate this consider a single-degree-of-freedom oscillator subjected to a base excitation whose equation of motion is

$$m\ddot{y}(t) + c\dot{x}(t) + kx(t) = 0 \qquad (5\text{-}39)$$

where

m = mass of oscillator = W/g

W = weight of oscillator

g = acceleration of gravity

c = damping coefficient

k = spring stiffness

$x(t)$, $\dot{x}(t)$ = oscillator mass's displacement and velocity relative to the support

$\ddot{y}(t)$ = absolute acceleration of the mass due to the earthquake support motion

If the damping term is neglected, then

$$\text{earthquake inertia force} = m\ddot{y}(t) = -kx(t) \qquad (5\text{-}40)$$

Therefore, the maximum force is equal to k times the maximum relative displacement induced by the earthquake excitation. Equation (5-26) relates this maximum displacement (S_d, or spectral displacement) to the spectral acceleration, and therefore

$$\text{maximum earthquake inertia force} = V = kS_d(T_n, \xi)$$

or

$$V = k \left(\frac{T_n}{2\pi}\right)^2 S_a(T_n, \xi) \qquad (5\text{-}41)$$

where

S_a = spectral acceleration

T_n = oscillator natural period of vibration in seconds

Recall that

ω_n^2 = natural frequency of vibration squared

$= k/m$ (radians/sec)^2

and

$$T_n = \frac{2\pi}{\omega_n}$$

Therefore,

$$\left(\frac{T_n}{2\pi}\right)^2 = \left(\frac{1}{\omega_n}\right)^2 = \frac{m}{k} = \frac{W}{kg}$$

and finally

$$V = \left(\frac{WS_a}{g}\right) \quad \text{or} \quad \frac{V}{W} = \frac{S_a}{g} \qquad (5\text{-}42)$$

Equation (5-42) shows that the ratio V/W is equal to the spectral acceleration when it is expressed in the units of g, not cm/sec^2 or in./sec^2. Figure 5-22 shows a plot of the spectral acceleration vs. natural period of vibration, and if the reader plots V/W from the UBC formula vs. T_n, it will be apparent that the general shape of the plot is very similar to the mean S_a plot of Figure 5-22.

The preceding comments of this section concern the calculation of lateral earthquake forces by using the equivalent static design approach. The other approach referred to at the start of this section was the *response spectrum earthquake design approach*. We cannot in this book go into depth in discussing the details of this latter approach because an adequate presentation requires that the reader be familiar with the theory of vibrations. However, it is possible, based on the material previously presented in this and the previous section, to provide a general treatment of the basic concept.

When a structure vibrates it is analytically possible to represent this vibrational motion as the superposition of motions associated with distinct deformation shapes. These shapes are called *mode shapes*. Associated with each mode shape is a natural period of vibration. The amplitude of motion in each mode shape is calculated by using the single-degree-of-freedom oscillator spectrum discussed in the previous section.

Recall that the response spectrum enables one to obtain a value of earthquake-induced spectral displacement, given the oscillator's natural period of vibration, damping, and mass. The amplitude of motion associated with any single mode shape is obtained by multiplying a spectral displacement value appropriate for that mode by a mass and mode shape-dependent constant called a *participation factor*. The appropriate spectral displacement value for the mode shape is obtained from the single-degree-of-freedom oscillator response spectrum with the oscillator period of vibration set equal to the mode shape natural period of vibration. The reason one needs to multiply this spectral displacement value by the participation factor is based on the analytical requirements of vibration theory. The total structural response is obtained by combining the lateral displacements calculated for each mode shape. There are different methods used by engineers for combining these modal displacements. For illustrative purposes one can imagine adding the absolute value of all modal displacements in order to obtain the total structural displacement.

In addition to calculating the deformations of the structure using mode shapes, one can calculate for each mode of vibration the lateral earthquake-induced inertia forces at all mass points (e.g., floor levels) and the base shear force. The resultant forces induced by the earthquake as calculated by using the response spectrum earthquake design approach are obtained by combining the forces associated with each of the vibrational mode shapes.

Several additional reading references are provided at the end of this chapter to assist the reader in further study.

Example 5-12

The earthquake base shear force for a building can be obtained from the UBC formula

$$V = ZIKCSW$$

Assume that the building is such that the following values apply

$$I = 1.0 \text{ (standard building)}$$

$$S = 1.0 \text{ (rock building site)}$$
$$Z = 1.0 \text{ (seismic zone 4)}$$

Plot the ratio V/W for $K = 0.67$, 1.0, and 1.33 vs. the building's fundamental period of vibration.

The ratio is equal to

$$\frac{V}{W} = KC$$

where

$$C = \frac{1}{15\sqrt{T_n}} \le 0.12$$

Figure Ex. 5-12 shows the V/W plot.

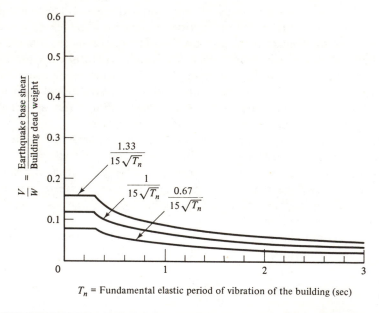

T_n = Fundamental elastic period of vibration of the building (sec)   *FIGURE Ex. 5-12*

Example 5-13

Consider the building discussed in Example 5-12. Assume that it has six stories of equal height and that its total height is 72 ft. Assume that $K = 1.33$ and that all floors are of equal weight.

Use the formula

$$T_n = \frac{0.05h_n}{\sqrt{D}}$$

where

$$D = \text{building dimension (in feet) in direction parallel to applied earthquake forces}$$
$$= 60 \text{ ft}$$

Calculate the lateral earthquake force at each floor level.

Since $h_n = 72$ ft and $D = 60$ ft, it follows that

$$T_n = \frac{0.05h_n}{\sqrt{D}} = \frac{0.05(72)}{\sqrt{60}} = 0.46 \text{ sec}$$

Therefore,

$$\frac{V}{W} = KC = \frac{1.33}{15\sqrt{T_n}} = 0.131$$

Note that $C = 0.098$ is less than 0.12, and therefore it is used in the calculation.
The concentrated load at the top floor, F_t, is zero because T_n is less than 0.7 sec.
The lateral earthquake force at the xth floor above the base of the building is

$$F_x = \left(\frac{wh_x}{\sum\limits_{i=1}^{6} wh_i}\right) V$$

where

$$w = \text{floor weight} = \frac{W}{6}$$

Note that the denominator in the floor force equation becomes

$$\sum_{i=1}^{6} wh_i = w(12 + 24 + 36 + 48 + 60 + 72) = 252w$$

Therefore,

$$F_x = \left(\frac{h_x}{252}\right) V = 0.131 \left(\frac{h_x}{252}\right) W$$

$$= 0.131 \left(\frac{h_x}{252}\right) (6w)$$

It follows that

$x = floor$	$F_x/w = earthquake\ force/floor\ weight$
1	0.037
2	0.075
3	0.112
4	0.150
5	0.187
6	0.225

The lateral earthquake force varies from 3.7 to 22.5% of the floor weight and is a maximum at the top floor. Some engineers like to visualize the lateral force in terms of acceleration units. Note that

$$\text{force} = \text{(mass)(acceleration)}$$

$$= \left(\frac{w}{g}\right)\text{(acceleration)}$$

Using the top floor as an example,

$$\text{force} = F_n = 0.225w = \left(\frac{w}{g}\right)\text{(acceleration)}$$

and finally

$$\text{acceleration} = 0.225g = 22.5\%\ \text{gravity}$$

Example 5-14

Consider Eq. (5-29); i.e.,

$$y = b_1 10^{b_2 M}(R + 25)^{-b_3}$$

where, from Table 5-11 (for $\xi = 0.05$ and $T_n = 0.5$ sec),

$$y = S_v(T, \xi) = \text{spectral velocity}$$
$$b_1 = 5.74$$
$$b_2 = 0.356$$
$$b_3 = 1.197$$

Assume that $M = 6.0$ and $R = 30$ km.

(a) Calculate the mean and standard deviation of S_v at $T_n = 0.50$ sec for $\xi = 0.05$.

    (b) Assume that T_n is deterministic, and calculate the mean and standard deviation of the corresponding spectral acceleration S_a.

    (c) Assume that S_a has a normal PDF. What is the probability that S_a will be greater than 0.131?

Part (a) solution:

$$\bar{y} = \text{mean spectral velocity}$$

$$= (5.74)(10^{(0.356)(6)})(30 + 25)^{-1.197} = 6.48 \text{ cm/sec}$$

$$\sigma_y = \text{standard deviation of spectral velocity (see Table 5-11)}$$

$$= (6.48)(0.591) = 3.831 \text{ cm/sec}$$

Part (b) solution:

$$S_a = \text{spectral acceleration} = \left(\frac{2\pi}{T_n}\right) S_v = \left(\frac{2\pi}{T_n}\right) y$$

Therefore,

$$\bar{S}_a = \text{mean of spectral acceleration} = \left(\frac{2\pi}{T_n}\right) \bar{y}$$

$$= 81.46 \text{ cm/sec}^2$$

and

$$\sigma_{S_a} = \text{standard deviation of spectral acceleration}$$

$$= \left(\frac{2\pi}{T_n}\right) \sigma_y = 48.14 \text{ cm/sec}^2$$

Part (c) solution:

The PDF of the random variable S_a is

$$p(S_a) = \frac{1}{\sigma_{S_a}\sqrt{2\pi}} \exp\left[-\frac{1}{2}\left(\frac{S_a - \bar{S}_a}{\sigma_{S_a}}\right)^2\right]$$

$$= \frac{1}{48.14\sqrt{2\pi}} \exp\left[-\frac{1}{2}\left(\frac{S_a - 81.46}{48.14}\right)^2\right]$$

It therefore follows, using Section 2-9, that

$$\Pr[S_a > 0.131g] = \Pr[S_a > 128.4 \text{ cm/sec}^2]$$

$$= \int_{128.4}^{\infty} p(S_a) dS_a$$

$$= 0.166, \text{ or } 16.6\%.$$

5-10 REFERENCES AND ADDITIONAL READING

References

[5.1] International Conference of Building Officials (1979): *Uniform Building Code*, Whittier, CA.

[5.2] CULVER, C.G. (1976): "Survey Results for Fire Loads and Live Loads in Office Buildings," *NBS Building Science Series Report 85*, Center for Building Technology, National Bureau of Standards, Washington, D.C.

[5.3] ROBERTSON, L.E., and T. NAKA (1980): "Tall Building Criteria and Loading," American Society of Civil Engineers, New York, Chapter CL-3.

[5.4] SIMIU, E. and R.H. SCANLAN (1978): *Wind Effects on Structures*, John Wiley & Sons, Inc., New York.

[5.5] THOM, H.C.S. (1968): "New Distribution of Extreme Winds in the United States," *Journal of the Structural Division*, ASCE, Vol. 94, ST 7.

[5.6] LEW, M. and G.C. HART (1979): "Microzonation for Wind," *Journal of the Structural Division*, ASCE, Vol. 105, ST 6.

[5.7] DAVENPORT, A.G. (1967): "Gust Loading Factors," *Journal of the Structural Division,* ASCE, Vol. 93, ST 3.

[5.8] American Society of Civil Engineers (1965): "Wind Forces on Structures," *Journal of the Structural Division*, ASCE, Vol. 91, ST 3.

[5.9] American National Standards Institute (1972): "American National Standard Building Code Requirements for Minimum Design Loads in Buildings and Other Structures," *ANSI A58.1-1972*.

[5.10] RICHTER, C.F. (1958): *Elementary Seismology*, W.H. Freeman & Company, Publishers, San Francisco.

[5.11] WEIGEL, R.L. (1970): *Earthquake Engineering*, Prentice-Hall, Inc., Englewood Cliffs, NJ.

[5.12] NEWMARK, N.M., and E. ROSENBLUETH (1971): *Fundamentals of Earthquake Engineering*, Prentice-Hall, Inc., Englewood Cliffs, NJ.

[5.13] HUDSON, D. (1979): *Strong Motion Record Analysis*, Earthquake Engineering Research Institute.

[5.14] ESTEVA, L. (1969): "Seismic Risk and Desimic Design Decisions," *Seminar on Seismic Design of Nuclear Power Plants*, The M.I.T. Press, Cambridge, MA.

[5.15] McGUIRE, R.K. (1974): "Seismic Structural Response Risk Analysis, Incorporating Peak Response Regressions on Earthquake Magnitude and Distance," Department of Civil Engineer Research Report R74-51, M.I.T.

[5.16] ROBERTSON, L.E., and T. NAKA (1980): "Tall Building Criteria and Loading," American Society of Civil Engineers, New York, Chapter CL-2.

[5.17] Applied Technology Council (1974): "An Evaluation of a Response Spectrum Approach to Seismic Design of Buildings," *ATC-2 Report*, San Francisco, CA.

Additional Reading

CLOUGH, R.W., and J. PENZIEN (1975): *Dynamics of Structures*, McGraw-Hill Book Company, New York.

MUTO, K. (1974): "A Seismic Design Analysis of Buildings," Maruzen Company, Ltd, Tokyo.

CHIOCEL, D., and D. LUNGU (1975): *Wind, Snow and Temperature Effects on Structures Based on Probability*," Abacus Press, Turnbridge Wells, Kent, England.

HOUSNER, G. (1947): "Characteristics of Strong-Motion Earthquakes," *Bull. Seism. Soc. Am.*, Vol. 37.

HOUSNER, G. (1959): "Behavior of Structures During Earthquakes," *Journal of the Engineering Mechanics Division*, ASCE, Vol. 85, EM 4.

CERMAK, J. (1975): "Applications of Fluid Mechanics to Wind Engineering," *ASME Journal of Fluids Engineering*, Vol. 97, No. 1.

HAUGEN, E.B. (1968): *Probabilistic Approaches to Design*, John Wiley & Sons, Inc., New York.

HOUGHTON, E.L., and N.B. CARRUTHERS (1976): *Wind Forces on Buildings and Structures*, John Wiley & Sons, Inc., New York.

MACDONALD, A.J. (1975): *Wind Loading on Buildings*, John Wiley & Sons, Inc., New York.

MEHTA, K. (1979): *Proceedings of the Workship on Wind Climate*, Asheville, NC.

BLEVINS, R.D. (1977): "*Flow-Induced Vibration*," Van Nostrand Reinhold Company, New York.

PROBLEMS

5.1 Consider the data shown in Figure 5-1. Assume that the PDF of the live load has a log-normal PDF with a mean and standard deviation equal to 9.0 psf and 5.5 psf, respectively.

    (a) Define the PDF and compare it with the data of Figure 5-1 by making a comparative sketch of the PDF and the data.

    (b) Use the PDF from part (a) and calculate the probability that the live load will be greater than 20 psf; greater than 45 psf. Compare your probabilities with the frequency data of Fig. 5-1.

5.2 A column supports a live load area of 300 sq ft. Assume that the dead load is 75 psf.

    (a) If the live load is obtained from Table 5-1 for the office category it is equal to 50 psf times a reduction percentage, denoted R. Calculate the value of R and the design live load.

    (b) Assume that the PDF of the live load magnitude is log-normal. Use the mean and standard deviation of the live load from Figure 5-3. and calculate the probability that the live load will exceed the design live load calculated in part (a).

5.3　The mean and coefficient of variation of the dead load D is given as 75 psf and 0.10, respectively. The mean and coefficient of variation of the live load L is given as 9.2 psf and 0.70, respectively. Assume that the total load T per square foot is

$$T = D + L$$

(a)　Calculate the mean and standard deviation of T if D and L are independent random variables.

(b)　Assume that D and L are independent normally distributed random variables. Calculate the probability that T will be greater than $\bar{D} + \bar{L}$ (i.e., $75.0 + 9.2 = 84.2$ psf).

(c)　Using the same PDF assumption as in part (b) calculate the probability that T will be greater than $1.2\bar{D} + \bar{L}$; greater than $\bar{D} + 1.2\bar{L}$; greater than $1.2\bar{D} + 1.2\bar{L}$.

5.4　Consider a simply supported steel beam with a deterministic section modulus equal to S. The beam is 20 ft long and it supports a uniform dead load D and a uniform live load L. The mean and coefficient of variation of the dead load is 750 lb/ft and 0.10, respectively. The mean and coefficient of variation of the live load is 92 lb/ft and 0.70, respectively. Assume that the dead and live loads are independent normally distributed random variables.

(a)　Calculate the mean and coefficient of variation of the load-induced moment at the midspan of the beam.

(b)　If the material yield stress is a normally distributed random variable, independent of D and L, with a mean equal to 47.9 ksi and a standard deviation equal to 3.3 ksi, then calculate the value of the section modulus S corresponding to a probability of failure equal to 0.05. Failure in this problem is defined to be when the load-induced moment is equal to or greater than the moment capacity of the cross section of the beam. The moment capacity is defined to be when the outside fiber of the cross section just reaches the yield stress.

(c)　Repeat part (b), but now assume that the moment capacity is the plastic moment capacity of the cross section and the section modulus S_p is the plastic section modulus.

5.5　The annual fastest-mile wind speed has been measured at a site for several years. The sample mean and sample coefficient of variation of the annual data is 70 mph and 0.20, respectively. Assume that the PDF of the annual fastest-mile wind speed is a Fréchet.

(a)　Calculate the parameters a and b in the Fréchet distribution function given by Eq. (5-4).

(b)　What is the probability that the fastest-mile wind speed for any single year will be greater than 70 mph? 80 mph?

(c) What is the probability that the fastest-mile wind speed for any single year will be greater than 70 mph but equal to or less than 80 mph?

5.6 Use the Fréchet PDF established in Problem 5.5.

(a) What is the mean return period for an annual fastest-mile wind speed greater than 80 mph?

(b) What is the mean return period for an annual fastest-mile wind speed greater than 70 mph but equal to or less than 80 mph?

(c) Calculate the value of the wind speed for a 50 year mean return period. This return period corresponds to a wind speed greater than the sought wind-speed value.

5.7 Fastest-mile wind-speed data has been obtained on a daily basis. The data for 10 consecutive days is as follows:

Day	Wind Speed (mph)
1	60.0
2	68.5
3	57.0
4	45.0
5	51.5
6	59.5
7	72.5
8	71.0
9	63.0
10	49.9

(a) Calculate the sample mean and variance and define the Fréchet PDF.

(b) What is the probability that the fastest-mile wind speed for any single day will be greater than 70 mph?

(c) What is the mean return period for a fastest-mile wind speed greater than 70 mph?

5.8 Snow load data has been obtained for a building site. Ten years of data exist, and for each year the data is the maximum annual snow load in pounds per square foot. The data is as follows:

Year	Snow Load (psf)
1	25.0
2	20.0
3	27.8
4	23.6
5	32.5
6	39.0
7	31.0
8	19.5
9	22.0
10	27.5

(a)   Calculate the sample mean and standard deviation of the data.

(b)   If the maximum annual snow load has a normal PDF, calculate the probability that the maximum annual snow load for any single year will be greater than 30 psf.

(c)   Calculate the mean return period for a maximum annual snow load greater than 30 psf.

5.9   Data exists for a site which indicates that the mean return period for the annual fastest-mile wind speed being greater than 100 mph is 100 years and for being greater than 90 mph is 80 years.

(a)   Calculate the corresponding values for a and b in Eq. (5-4), and sketch the probability distribution function.

(b)   If the mean return period is 50 years for a wind speed being greater than V_0, calculate V_0.

5.10 The probability of the annual fastest-mile wind speed exceeding 80 mph at an elevation of 10 m for an open-country site is 0.05. Assume that

$$H_G = \text{open-country gradient height} = 275 \text{ m}$$

$$\alpha = \text{open-country power law coefficient} = 0.16$$

Assume that the power law profile applies.

(a)   Calculate the value of the corresponding wind speed at the gradient height (i.e., V_G).

(b)   Sketch the corresponding vertical variation of wind speed.

(c)   What is the probability that the annual fastest-mile wind speed at the gradient height will be greater than the V_G you calculated in part (a)?

5.11 Assume that the V_G you calculated in Problem 5.10 is the same for a site located near the original open-country site. The new site has a suburban exposure, and therefore

$$H_G = 395 \text{ m}$$

$$\alpha = 0.28$$

(a)   Calculate the annual fastest-mile wind speed at a height of 10 m for this new site and for the calculated V_G from Problem 5.10.

(b)   Sketch the vertical variation of the corresponding annual fastest-mile wind speed from the ground level to a height of 20 m.

5.12 Repeat Problem 5.11, but now assume that the new site has a large city or urban exposure with

$$H_G = 520 \text{ m}$$

$$\alpha = 0.40$$

Compare the value of the annual fastest-mile wind speed at a height of 20 m from Problems 5.10, 5.11, and 5.12. Which exposure results in the greatest mean annual fastest-mile wind speed at this height? Is the conclusion only valid for a height of 20 m?

5.13 Consider the structure shown in Fig. 5-13 with the h, b, and l ratios of 1.0, 4.0, and 4.0, respectively. Assume the mass density of air is

$$\rho = 0.00512 \ (\text{lb/ft}^2)/(\text{mph})^2$$

Assume that the wind speed at the height h is equal to 60 mph and that $h = 30$ ft. Let the internal pressure coefficient be ± 0.30 and $G = 1.2$.

(a) Calculate the dynamic pressure per unit area at $h = 30$ ft.

(b) If the angle of wind attack (ϕ) is $0°$, calculate the mean wind pressure per unit area for each of the zones E to H on the roof and A to D on the faces. Make a sketch which indicates the directions of the pressures.

(c) Compare your pressures in part (b) with those for an angle of wind attack ϕ of $45°$.

Note that all pressures act normal to the structural surface.

5.14 Repeat Problem 5.13, but now with h, b, and l ratios of 1.0, 1.0, and 1.0 respectively. Assume that $h = 30$ ft.

5.15 Repeat Problem 5.13, but now with h, b, and l ratios of 2.5, 1.0, and 1.0, respectively. Assume that $h = 30$ ft.

5.16 The structure in Problem 5.13 is to have a glass skylight on the roof, located in zone G. Use the same wind speed, i.e. 60 mph, and determine the wind pressures on the window for the three angles of wind attack $\phi = 0°$, $45°$, and $90°$. Assume that the internal pressure coefficient (C_{pi}) is equal to 0.3. If a gust factor G is used and it were equal to 1.2, how would your answers change?

5.17 The equation for the wind pressure on a particular building window is

$$p = (C_e + C_i)G(\tfrac{1}{2}\rho V^2)$$

where

p = pressure per unit area (lb/ft^2)

C_e = external pressure coefficient

C_i = internal pressure coefficient

G = gust factor

ρ = mass density of air = 0.00512 (lb/ft^2)/(mph)^2

V = annual fastest-mile wind speed

Consider p to be deterministic and C_e, C_i, G, and V to be random variables. Assume the following:

$$\bar{C}_e = \text{mean of } C_e = 0.90$$

$$\sigma_{C_e} = \text{standard deviation of } C_e = 0.18$$

$$\bar{C}_i = \text{mean of } C_i = 0.30$$

$$\sigma_{C_i} = \text{standard deviation of } C_i = 0.09$$

$$\bar{G} = \text{mean of } G = 1.20$$

$$\sigma_G = \text{standard deviation of } G = 0.24$$

$$\bar{V} = \text{mean of } V = 60.0 \text{ mph}$$

$$\sigma_V = \text{standard deviation of } V = 18.0 \text{ mph}$$

(a) Use a linear statistical analysis and calculate the mean and standard deviation of the window pressure.

(b) If the window pressure is assumed to have a log-normal PDF, then define and sketch the PDF of the load-induced pressure.

(c) Assume that the window thickness is such that it has a resistance against wind pressure (denoted R) with a log-normal PDF. The mean and coefficient of variation of R are 20.0 psf and 20%, respectively. Define and sketch the PDF window load resistance pressure.

(d) Use the results from Section 4-2 and calculate the probability of window failure.

5.18 Problem 5.17 uses a linear statistical analysis to analyze a particular building window. Now assume that C_e, C_i, G, and V are random variables with the mean and standard deviations specified in Problem 5.17. For computational convenience assume that each of these random variables is independent and has uniform probability density functions.

(a) Use the random numbers from Table 3-1 and calculate the mean and standard deviation of the pressure. Plot the histogram of the pressure. (Note: Use column one of random numbers for C_e, the column two for C_i, etc.) Use only the first 20 numbers in each column in the Monte Carlo study. Recall that the numbers in the table are for a uniform 0 to 1 PDF and must be transformed prior to being used in the pressure equation.

(b) Discuss how you would extend the Monte Carlo solution in part (a) in order to calculate the probability of window failure.

5.19 Consider an earthquake with a Richter magnitude of 6.0. Calculate:
(a) The energy release.
(b) The relative fault displacement.
(c) The length of fault rupture.

5.20 Repeat Problem 5.19, but now use a Richter magnitude of 8.0.

5.21 Calculate the peak ground velocity for MMI values VII, IX, and XI.

5.22 A system has 5% damping and a natural period of vibration of 0.5 sec. The spectral velocity is given and is equal to 20 cm/sec. Calculate:

(a) The spectral displacement of system response.

(b) The spectral acceleration of system response.

5.23 Use Figure 5-20(a) and calculate the spectral velocity for a system with a natural period of vibration and damping equal to 1.0 sec and 2%, respectively.

5.24 Use Figure 5-20(b) and calculate the spectral displacement, velocity, and acceleration for a system with a natural period of vibration and damping equal to 0.3 sec and 2%, respectively.

5.25 Consider the historical earthquake data for the time period 1960 to 1979 as shown in the following table. Only events of Richter magnitude 5.0 or greater are considered. The hypocentral distance is to a selected building site.

Event Number	Richter Magnitude	Hypocentral Distance (km)
1	6.0	25
2	6.5	40
3	5.5	30
4	5.2	15
5	7.0	35

(a) Calculate the number of earthquakes per year with an $M = 5.0$ or greater. What is the corresponding return period?

(b) Fit the data, using the least-squares fit technique, with the equation

$$\log N = a - bM$$

where

$$N = \text{number of earthquakes per year with a}$$
$$\text{magnitude equal to or greater than } M$$

$$a, b = \text{parameters determined from the least-squares fit}$$

(c) Calculate the number of earthquakes per year with an $M = 5.0$ or greater, using the equation established in part (b). How does this estimate compare with your result from part (a)?

(d) Calculate the number of earthquakes per year with an $M = 7.5$ or greater, using the equation established in part (b). Why is it often necessary to extrapolate beyond the actual data and why can this be dangerous?

5.26 Use your results from Problem 5.25, part (b), and calculate:

(a) The return period for earthquakes with a Richter magnitude in the range $6.0 \leq M < 6.5$.

(b) Repeat part (a), but with the range now equal to $6.5 \leq M < 7.0$.

5.27 Use the historical data of Problem 5.25 and calculate, using Eq. (5-29), the following:

(a)   The mean of the peak ground acceleration.

(b)   The standard deviation and coefficient of variation of the peak ground acceleration.

5.28   Consider a postulate earthquake of Richter magnitude 8.0 at a hypocentral distance of 40 km for a building site. Calculate, using Table 5-11, the mean and standard deviation of the following quantities:

(a)   Peak ground acceleration.

(b)   Spectral velocity of an oscillator with a natural period of 1 sec and 5% damping.

If each of these response quantities is assumed to have a normal PDF, then plot their PDF and also calculate the value of each response quantity which corresponds to a 10% probability of non-exceedance.

5.29   A building has 10 stories. Each story has a height of 13 ft, and therefore the total building height is 130 ft. Assume that the dead weight of each floor is w. The width of the building in the direction of the earthquake loading is 50 ft. Let

$$I = 1.00$$

$$Z = 1.00$$

$$K = 0.67$$

$$T_s = 1.00 \text{ sec}$$

Calculate the following:

(a)   The earthquake base shear in terms of w.

(b)   The lateral earthquake force at each floor level in terms of w.

(c)   The spectral acceleration of the system response, using Figure 5-20(a).

5.30   Consider the building described in Problem 5.29. Now assume that the building has only five stories. Repeat parts (a), (b), and (c) of Problem 5.29, assuming that all other given information remains unchanged. Compare your results for these two buildings of different heights.

5.31   A building has been designed such that the UBC earthquake base shear is 15% of the building's dead weight. Therefore, $V/W = 0.15 \sim S_a^d$, where S_a^d is the design spectral acceleration and its units are in g's.

(a)   Calculate the mean and standard deviation of the random variable spectral acceleration S_a for a Richter magnitude 8.0 earthquake at a hypocentral distance of 40 km from the building site. Assume $T_n = 0.5$ sec and 5% damping.

(b)   Assume that S_a from part (a) has a normal PDF. Plot the PDF.

(c)   Define the following damage states:

Light damage when $1.0 S_a^d \leq S_a \leq 2.0 S_a^d$.

Major damage when $2.0 S_a^d < S_a \leq 3.0 S_a^d$.

Collapse when $S_a > 3.0 S_a^d$.

Calculate the probability of each damage state occurring for the earthquake defined in parts (a) and (b). Use your plot from part (b) to describe the damage ranges.

(d) In part (c) the upper and lower limits of the damage state definitions are considered to be deterministic. Discuss how you would use a Monte Carlo analysis to solve the problem if these limits are considered to be random variables.

Answers to Selected Problems

CHAPTER 2

2.1   (a) Mean = 3.51×10^6 psi
           Variance = 7.83×10^9 psi
           Coefficient of variation = 2.52×10^{-2}
    (b) Mean = 7.26×10^3 psi
           Variance = 9.01×10^5 psi
           Coefficient of variation = 1.31×10^{-1}
    (c) Covariance = 5.62×10^7 psi
           Correlation coefficient = 6.69×10^{-1}

2.4   (a) 80%
    (b) 20%
    (c) 70%
    (d) 20%

2.5   (b) $C_o = 1.11 \times 10^{-11}$/psi
    (c) 5.55%
    (d) 72.15%

2.8   Mean $= \dfrac{y}{I}\,\overline{M}$

       Variance $= \dfrac{y^2}{I^2}\,\sigma M^2$

Coefficient of variation $= \dfrac{\sigma M}{\overline{M}}$

2.10 (a) Mean $= \dfrac{l}{2}\overline{P} + \dfrac{l^2}{2}\overline{W}$

Standard deviation $= \sqrt{\dfrac{l^2}{4}\sigma_p^2 + \dfrac{l^4}{4}\sigma w^2}$

Coefficient of variation $= \dfrac{\sqrt{\sigma_p^2 + l^2 \sigma w^2}}{\overline{P} + l\overline{w}}$

(b) Mean $= \dfrac{l}{2}\overline{P} + \dfrac{l^2}{2}\overline{W}$

Standard deviation $= \sqrt{\dfrac{l^2}{4}\sigma_p^2 + \dfrac{l^4}{4}\sigma_w^2 + \dfrac{l^6}{8}\rho\sigma_p\sigma_w}$

Coefficient of variation $= \dfrac{\sqrt{\sigma_p^2 + l^2 \sigma w^2 + l^4 \rho\sigma\sigma w / 2}}{\overline{P} + l\overline{w}}$

2.11 $\overline{X}_1 = 3.5 \times 10^6$ psi
$\rho_{x1} = 0.062$

2.17 (b) 0.073
(c) 0.23
(d) 0.54

2.19 (a) $\{Y\} = \begin{Bmatrix} M \\ \Delta \end{Bmatrix}$ $\{x\} = \begin{Bmatrix} P \\ W \end{Bmatrix}$ $[c] = \begin{bmatrix} \dfrac{l}{8} & \dfrac{l^2}{12} \\ \dfrac{l^3}{192EI} & \dfrac{l^4}{384EI} \end{bmatrix}$

(b) $\dfrac{\overline{M}}{\overline{\Delta}} = \dfrac{12.5\,\overline{P} + 833.33\,\overline{W}}{1.736 \times 10^{-6}\,\overline{P} + 8.681 \times 10^{-5}\,\overline{W}}$

$$[Sy] = \begin{bmatrix} Sy_{11} & Sy_{12} \\ Sy_{21} & Sy_{22} \end{bmatrix}$$

$Sy_{11} = 156.25 \text{ Var} (M) + 20833 \text{ Cov} (M, \Delta) + 6.94 \times 10^5 \text{ Var} (\Delta)$

$Sy_{12} = 2.08 \times 10^{-5} \text{ Var} (M) + 2.53 \times 10^{-3} \text{ Cov} (M, \Delta) + 7.23 \times 10^{-2}$
$\text{Var} (\Delta)$

$Sy_{22} = 3.01 \times 10^{-12} \text{ Var} (M) + 3.08 \times 10^{-10} \text{ Cov} (M, \Delta) + 754 \times 10^{-9} \text{ Var} (\Delta)$

CHAPTER 3

3.1 (a) $\Delta = \dfrac{\bar{P} l^3}{3 EI} + \dfrac{l^3}{3 EI} \quad (P - \bar{P})$

(c) $M = \dfrac{\bar{P} l}{8} + \dfrac{\bar{W} l^2}{12} + \dfrac{l}{8} \quad (P - \bar{P}) + \dfrac{l^2}{12} \quad (W - \bar{W})$

3.6 $\bar{M} = \dfrac{\bar{P} l}{8} + \dfrac{\bar{W} l^2}{12}$

$\sigma M = \dfrac{l^2}{64} \sigma_{P^2} + \sqrt{\dfrac{l^3}{48}} \text{ Cov } (P, W) + \dfrac{l^4}{144} \sigma_W{}^2$

3.9 (a) $\bar{R}_A = \bar{R}_E = \dfrac{\bar{P}}{2}$

$\sigma_{R_A} = \sigma R_E = \dfrac{1}{2} \sigma_P$

(b) $\bar{F}_{AF} = \dfrac{5}{6} \bar{P}$

$\sigma_{AF} = \dfrac{5}{6} \sigma_P$

3.12 (a) Mean = 4.83 ksi
Standard deviation = 0.757 ksi

3.16

Sample	fc' (psi)
1	7418
2	8737
3	9397
4	7614
5	7719
6	8221
7	8499
8	5605

3.17	Sample	fc' (psi)	*(continued)*
	9	9230	
	10	5212	
	11	6512	
	12	6920	
	13	8531	
	14	9264	
	15	5087	
	16	5035	
	17	7571	
	18	8480	
	19	6152	
	20	9071	

3.18	Sample	R
	1	1.0360
	2	1.1373
	3	1.2292
	4	1.0494
	5	1.0566
	6	1.0920
	7	1.1156
	8	0.8911
	9	1.1974
	10	0.8290
	11	0.9729
	12	1.0021
	13	1.1184
	14	1.2029
	15	0.7940
	16	0.7724
	17	1.0464
	18	1.1140
	19	0.9447
	20	1.1743

3.20	Sample	fc' (psi)
	1	6.731
	2	7.795
	3	10.233
	4	6.862
	5	6.932
	6	7.306
	7	7.563
	8	5.431

3.20 | Sample | fc' (psi) | *(continued)* |
|---|---|---|
| 9 | 8.726 | |
| 10 | 4.540 | |
| 11 | 6.145 | |
| 12 | 6.411 | |
| 13 | 7.594 | |
| 14 | 9.026 | |
| 15 | 4.232 | |
| 16 | 3.881 | |
| 17 | 6.833 | |
| 18 | 7.546 | |
| 19 | 5.894 | |
| 20 | 8.271 | |

3.21 (a) | Sample | E (psi) |
|---|---|
| 1 | 3.53×10^6 |
| 2 | 3.68×10^6 |
| 3 | 3.75×10^6 |
| 4 | 3.55×10^6 |
| 5 | 3.56×10^6 |
| 6 | 3.62×10^6 |
| 7 | 3.65×10^6 |
| 8 | 3.33×10^6 |
| 9 | 3.73×10^6 |
| 10 | 3.29×10^6 |
| 11 | 3.43×10^6 |
| 12 | 3.48×10^6 |
| 13 | 3.65×10^6 |
| 14 | 3.73×10^6 |
| 15 | 3.28×10^6 |
| 16 | 3.27×10^6 |
| 17 | 3.55×10^6 |
| 18 | 3.65×10^6 |
| 19 | 3.39×10^6 |
| 20 | 3.71×10^6 |

(b) Mean = 2.83×10^{-7} psi^{-1}
Variance = 1.64×10^{-16} psi^{-2}
(c) Mean = 2.84×10^{-7} psi^{-1}
Variance = 1.43×10^{-16} psi^{-2}

CHAPTER 4

4.2 Probability of failure = 19.5%

4.5 (a) $A = 2.20$ in^2

(b) $A = 2.67$ in^2

4.7   Probability of failure = 20.7%

4.10 Probability of failure = 21.8%

CHAPTER 5

5.2   (a) $R = 24\%$
Design live load = 38 psf
(b) 0.6%

5.3   (a) $\overline{T} = 84.2$ psf
$\sigma_T = 9.89$ psf
(b) 50%
(c) 6.4%
42.8%
4.5%

5.7   (a) $\overline{V} = 59.8$ mph

$\sigma_V{}^2 = 76.6$ mph.

$$P(V) = \left[\frac{10}{55.94}\right] \left[\frac{55.94}{V}\right]^9 \exp\left\{-\left[\frac{55.94}{V}\right]^{10}\right\}$$

(b) 10.1%
(c) 9.9 days

5.10 (a) $V_G = 136$ mph
(c) 5%

5.13 (a) 9.22 psf
(b)

Zone	Pressure (psf)
A	13.28
B	− 6.64
C	− 7.74
D	− 7.74
E	−12.17
F	−12.17
G	− 6.64
H	− 6.64

(c)

Zone	Pressure (psf)
A	8.85
B	− 7.74
C	8.85

5.13 (c)

Zone	Pressure (psf)	*(continued)*
D	– 7.74	
E	–13.28	
F	– 9.96	
G	– 9.96	
H	– 6.64	

5.17 (a) $P = 13.27$ psf
$\sigma p = 8.59$ psf

(b) $P(P) \dfrac{1}{1.484\ p}\ \exp\left\{-\dfrac{1}{2}\left[\dfrac{\ln p - 2.410}{0.592}\right]^2\right\}$

(c) $p\ (R) = \dfrac{1}{0.496\ R}\ \exp\left\{-\dfrac{1}{2}\left[\dfrac{\ln R - 2.975}{0.198}\right]\right\}$

(d) Probability of failure = 23.9%

5.18 (a) $P = 13.33$ psf
$\sigma p = 7.64$ psf

5.19 (a) $W = 6.3 \times 10^{20}$ ergs
(b) $D = 1.07$ ft
(c) $L = 60.1$ miles

5.24 $S_d = 1.00$ cm
$S_v = 51$ cm/sec
$S_a = 1.2$ g

5.29 (a) $V = 0.60\ W$

(b)

Floor	Lateral Force
Roof	0.141 W
10	0.092 W
9	0.082 W
8	0.071 W
7	0.061 W
6	0.051 W
5	0.041 W
4	0.031 W
3	0.020 W
2	0.010 W
Base	0.60 W

(c) $S_a = 21.9$ ft/sec^2

Index